Book series on industrial innovation

Series editor

Adedeji B. Badiru
Department of Systems and Engineering Management
Air Force Institute of Technology (AFIT) – Dayton, Ohio

Published titles:

Handbook of Industrial and Systems Engineering
Adedeji B. Badiru

Techonomics: The Theory of Industrial Evolution
H. Lee Martin

Forthcoming titles:

Computational Economic Analysis for Engineering and Industry
Adedeji B. Badiru & Olufemi A. Omitaomu

Industrial Project Management: Concepts, Tools and Techniques
Adedeji B. Badiru, Abi Badiru, & Ade Badiru

Beyond Lean: Elements of a Successful Implementation
Rupy (Rapinder) Sawhney

Triple C Model of Project Management: Communication, Cooperation, Coordination
Adedeji B. Badiru

Process Optimization for Industrial Quality Improvement
Ekepre Charles-Owaba & Adedeji B. Badiru

Systems Thinking: Coping with 21st Century Problems
John Turner Boardman & Brian J. Sauser

Computational Economic Analysis for Engineering and Industry

Computational Economic Analysis for Engineering and Industry

ADEDEJI B. BADIRU

OLUFEMI A. OMITAOMU

CRC Press
Taylor & Francis Group
Boca Raton   London   New York

CRC Press is an imprint of the
Taylor & Francis Group, an **informa** business

CRC Press
Taylor & Francis Group
6000 Broken Sound Parkway NW, Suite 300
Boca Raton, FL 33487-2742

International Standard Book Number-13: 978-0-8493-7477-7 (Hardcover)

Visit the Taylor & Francis Web site at
http://www.taylorandfrancis.com

and the CRC Press Web site at
http://www.crcpress.com

Acknowledgments

We thank all those who contributed to the completion and quality of this book. We thank our students at the University of Tennessee for their insights and ideas for what to include and what to delete. We especially thank Sirisha Saripali-Nukala and Bukola Ojemakinde for their help in composing many of the examples and problems included in the book. We also thank Jeanette Myers and Christine Tidwell for their manuscript preparation and administrative support services throughout the writing project. Many thanks go to Em Chitty Turner for her expert editorial refinement of the raw manuscript. We owe a debt of gratitude to Cindy Renee Carelli, senior acquisitions editor at CRC Press, and her colleagues for the excellent editorial and production support they provided for moving the book from idea to reality.

Dedication

This book is dedicated to our families, from whom we directed our attention occasionally while we labored on this manuscript.

Table of Contents

Preface

The bottom line, expressed in terms of cost and profits, is a major concern of many organizations. Even public institutions that have traditionally been nonchalant about costs and profits are now beginning to worry about economic justification. Formal economic analysis is the only reliable mechanism through which all the cost ramifications of a project, public or private, can be evaluated. Good economic analysis forms the basis for good decision making. Modern decisions should have sound economic basis. Decisions that don't have good economic foundations will eventually come back to haunt the decision maker. This is true for all types of decisions including technology decisions, engineering decisions, manufacturing decisions, and even social and political decisions.

Decision making is the principal function of an engineer or manager. Over two-thirds of engineers will spend over two-thirds of their careers as managers and decision makers. Ordinary decision making involves choosing between alternatives. Economic decision making involves choosing between alternatives on the basis of monetary criteria. Economic analysis is a fundamental tool of the decision-making process. Traditionally, the application of economic analysis techniques to engineering problems has been referred to as engineering economy or engineering economic analysis. However, the increasing interest in economic analysis in all disciplines has necessitated a greater use of the more general term, economic analysis. Over the past few decades, the interest in engineering economy has increased dramatically. This is mainly due to a greater awareness and consciousness of the cost aspects of projects and systems. In the management of technology, it is common for top management people to consist largely of former engineers. It has, consequently, become important to train engineers in the cost aspects of managing engineering and manufacturing systems. Colleges and universities have been responding to this challenge by incorporating engineering economy into their curricula. All engineering and technology disciplines now embrace the study of engineering economy. This has, consequently, created a big and delineated market with a growing demand for engineering economy books. The rapid development of new engineering technologies and the pressure to optimize systems output while minimizing cost has continued to fuel the market momentum. Unfortunately, the pace of generating text materials for engineering economy has not kept up with the demand. This is even more so in the very complex industrial environment, where integrated computational analyses are required.

This book on **Computational Economic Analysis for Engineering and Industry** provides direct computational tools, techniques, models, and approaches for economic analysis with a specific focus on industrial and engineering processes. The book integrates mathematical models, optimization, computer analysis, and the managerial decision process. Industry is a very dynamic and expansive part of the national economy that is subject to high levels of investment, risks, and potential economic rewards. To justify the investments, special computational techniques must be used to address the various factors involved in an industrial process.

A focused compilation of formulations, derivations, and analyses that have been found useful in various economic analysis applications will be of great help to industry professionals. This book responds to the changing economic environment of industry. Recent global economic anxiety indicates that more focus needs to be directed at economic issues related to industry. The book provides a high-level technical presentation of economic analysis of the unique aspects of industrial processes. Existing conventional techniques, while well proven, do not adequately embrace the integrated global factors affecting unique industries. Conceptual and philosophical publications are available on the worldwide developments in industry. But industry-focused computational tools are not readily available. This book fills that void. The contents of the book include new topics such as:

New economic analysis models and techniques
Tent-shaped cash flows
Industrial economic analysis
Project-based economic measures
Profit ratio analysis
Equity break-even point
Utility based analysis
Project-balance analysis
Customized *ENGINEA* software tool

The book will provide students, researchers, and practitioners with a comprehensive treatment of economic analysis, considering the specific needs of industry. Topics such as investment justification, break-even analysis, and replacement analysis are covered in an updated manner. The book provides a pragmatic alternative to conventional economic analysis books. Readers will find useful general information in the Appendixes, which contain engineering conversion factors and formulae.

Adedeji B. Badiru
Olufemi A. Omitaomu
2007

The Authors

Adedeji "Deji" B. Badiru is the department head of systems and engineering management at the U.S. Air Force Institute of Technology (AFIT), Wright Patterson Air Force Base, Ohio. Previously head of industrial and information engineering at the University of Tennessee in Knoxville, he served as professor of industrial engineering and dean of University College at the University of Oklahoma. He is a registered professional engineer, a fellow of the Institute of Industrial Engineers, and a fellow of the Nigerian Academy of Engineering. He holds a B.S. degree in industrial engineering, an M.S. in mathematics, an M.S. in industrial engineering from Tennessee Technological University, and a Ph.D. in industrial engineering from the University of Central Florida.

His areas of expertise and courses taught cover mathematical modeling, project management, systems analysis, and economic analysis. He is the author of several technical papers and books, and is the editor of the *Handbook of Industrial and Systems Engineering*. He is a member of several professional associations, including the Institute of Industrial Engineers (IIE), Society of Manufacturing Engineers (SME), Institute for Operations Research and Management Science (INFORMS), American Society for Engineering Education (ASEE), American Society for Engineering Management (ASEM), and the Project Management Institute (PMI).

He has served as a consultant to several organizations around the world, including Russia, Mexico, Taiwan, Venezuela, South Africa, Nigeria, Ghana and South Korea. He has conducted customized training workshops for numerous organizations including Sony, AT&T, Seagate Technology, the U.S. Air Force, Oklahoma Gas and Electric, Oklahoma Asphalt Pavement Association, Hitachi, Nigeria National Petroleum Corporation, and ExxonMobil. He is the recipient of several honors including the IIE Outstanding Publication Award, University of Oklahoma Regents' Award for Superior Teaching, School of Industrial Engineering Outstanding Professor of the Year, Eugene L. Grant Award for Best Paper in Volume 38 of *The Engineering Economist* journal, University of Oklahoma College of Engineering Outstanding Professor of the Year, Ralph R. Teetor Educational Award from the Society of Automotive Engineers, Award of Excellence as chapter president from the Institute of Industrial Engineers, UPS Professional Excellence Award, Distinguished Alumni Award from Saint Finbarr's College, Lagos, Nigeria, and Distinguished

Alumni Award from the Department of Industrial and Systems Engineering, Tennessee Tech University. He holds a leadership certificate from the University of Tennessee Leadership Institute.

Dr. Badiru has served as a technical project reviewer for the Third-World Network of Scientific Organizations, Italy. He has also served as a proposal review panelist for the National Science Foundation and National Research Council, and a curriculum reviewer for the American Council on Education. He is on the editorial and review boards of several technical journals and book publishers, and was an industrial development consultant to the United Nations Development Program.

Olufemi Abayomi Omitaomu received a B.S. degree in mechanical engineering from Lagos State University, Nigeria in 1995, an M.S. degree in mechanical engineering from the University of Lagos, Nigeria in 1999, and a Ph.D. in industrial engineering from University of Tennessee, Knoxville, in 2006. He won several academic prizes during his undergraduate and graduate programs. After his B.S., he worked as a project engineer for Mobil Producing Nigeria between 1995 and 2001. During his Ph.D. program, he taught engineering economic analysis course for several semesters. Olufemi has published several journal and conference articles in international journals including *The Engineering Economist*. He has also published book chapters on economic analysis and data mining techniques. He jointly published two computer software programs including the ENGINEA, which is included in this book. Olufemi is a member of several professional bodies including the Institute of Industrial Engineers (IIE), Institute of Electrical and Electronic Engineers (IEEE), Institute for Operations Research and the Management Sciences (INFORMS), and the American Society of Mechanical Engineers (ASME). He was a board member of the Engineering Economy Division, Institute of Industrial Engineers (IIE) for 4 years. He is listed in the 2006 edition of *Who's Who in America* and 2007 edition of *Who's Who in Science and Engineering*. He is currently a research associate in the Computational Sciences and Engineering Division at Oak Ridge National Laboratory, Tennessee. His research interests include computational economic analysis and online knowledge discovery and data mining. He is married to Remilekun Enitan, and they have two children, Oluwadamilola and Oluwatimilehin.

December 2006

chapter one

Applied economic analysis

Industrial enterprises have fundamentally unique characteristics and require unique techniques of economic analysis. Thus, although the methodologies themselves may be standard, the specific factors or considerations may be industrially focused. Fortunately, most of the definitions used in general economic analyses are applicable to industrial economic analysis. This chapter presents computational definitions, techniques, and procedures for applied industrial economic analysis. As in the chapters that follow, a project basis is used for most of the presentations in this chapter.

1.1   Cost- and value-related definitions

We need to define and clarify some basic terms often encountered in economic analysis. Some terms appear to be the same but are operationally different. For example, the term *economics* must be distinguished from the term *economic analysis*, and even more specifically from the term *engineering economic analysis*. *Economics* is the study of the allocation of the scarce assets of production for the purpose of satisfying some of the needs of a society. *Economic analysis*, in contrast, is an integrated analysis of the qualitative and quantitative factors that influence decisions related to economics. Finally, an *engineering economic analysis* is an analysis that focuses on the engineering aspects. Examples of the engineering aspects typically considered in an economic design process include the following:

- Product conceptualization
- Research and development
- Design and implementation
- Prototyping and testing
- Production
- Transportation and delivery

Industrial economics is the study of the relationships between industries and markets with respect to prevailing market conditions, firm behavior, and economic performance. In a broader sense, the discipline of industrial economics focuses on a broad mix of industrial operations involving real-world competition, market scenarios, product conceptualization, process development, design, pricing, advertising, supply chain, delivery, investment strategies, and so on. Although this book may touch on some of the cost aspects, the full range of industrial economics is beyond its scope. Instead, the book focuses on the computational techniques that are applied to economic analysis in industrial settings.

Earned value analysis is often used in industrial project economic analysis to convey the economic status of a project. *Planned value* (PV) refers to the portion of the approved cost that is planned to be spent during a specific period of the project. *Actual cost* (AC) is the total direct and indirect costs incurred in accomplishing work over a specific period of time. *Earned value* (EV) is defined as the budget for the work accomplished in a given period. Formulas relating to these measures are used to assess the overall economic performance of a project. Specific definitions are presented below:

- *Cost variance* (CV) equals EV minus AC. The cost variance at the end of the project is the difference between the *budget at completion* (BAC) and the actual amount spent:

$$CV = EV - AC$$

  A positive CV value indicates that costs are below budget.
  A negative CV value indicates a cost overrun.
- *Schedule variance* (SV) equals EV minus PV. Schedule variance will ultimately equal zero when the project is completed, because all of the planned values will have been earned:

$$SV = EV - PV$$

  A positive SV value indicates that a project is ahead of schedule.
  A negative SV value indicates that the project is behind schedule.
- *Cost performance index* (CPI) equals the ratio of EV to AC. A CPI value less than 1.0 indicates a cost overrun of estimates. A CPI value greater than 1.0 indicates a cost underrun of estimates. CPI is the most commonly used cost-efficiency indicator:

$$CPI = EV/AC$$

  A CPI greater than 1.0 indicates costs are below budget.
  A CPI less than 1.0 indicates costs are over budget.
- *Cumulative CPI* (CPI^C) is used to forecast project costs at completion. CPI^C equals the sum of the *periodic earned values* (EV^C) divided by the sum of the individual *actual costs* to date (AC^C):

$$CPI^C = EV/AC^C$$

- A *schedule performance index* (SPI) is used, in addition to the schedule status, to predict completion date and is sometimes used in conjunction with CPI to generate project completion estimates. SPI equals the ratio of EV to PV:

$$SPI = EV/PV$$

An SPI greater than 1.0 indicates that a project is ahead of schedule. An SPI less than 1.0 indicates that a project is behind schedule.
- BAC, AC^C, and *cumulative cost performance index* (CPIC) are used to calculate the *estimated total cost* (ETC) and the *estimated actual cost* (EAC), where BAC is equal to the total PV at completion for a scheduled activity, work package, control account, or other WBS component:

$$BAC = \text{total cumulative PV at completion}$$

- ETC, based on atypical variances, is an approach that is often used when current variances are seen as atypical and the project management team expects that similar variances will not occur in the future. ETC equals BAC minus the *cumulative earned value* to date (EV^C):

$$ETC = (BAC - EV^C)$$

- *Estimate at completion* (EAC), also used interchangeably with estimated actual cost, is the expected total project cost upon completion with respect to the present time. There are alternate formulas for computing EAC depending on different scenarios. In one option, EAC equals AC^C plus a new ETC that is provided by the project organization. This approach is most often used when past performance shows that the original estimating assumptions are no longer applicable due to a change in conditions:

$$EAC = AC^C + ETC$$

- *EAC using remaining budget.* EAC equals AC^C plus the budget required to complete the remaining work, which is BAC minus EV. This approach is most often used when current variances are seen as atypical and the project management team expects that similar variances will not occur in the future:

$$EAC = AC^C + BAC - EV$$

- *EAC using CPI^C.* EAC equals AC^C to date plus the budget required to complete the remaining project work, which is BAC minus EV, modified by a performance factor (often CPI^C). This approach is most often used when current variances are seen as typical of future variances:

$$EAC = AC^C + ((BAC - EV)/CPI^C)$$

- *Present Value* (PV) is the current value of a given future cash-flow stream, discounted at a given rate. The formula for calculating a present value is:

$$PV = FV/(1+r)^{(n)}$$

1.2   *Economics of worker assignment*

Operations research techniques are often used to enhance resource allocation decisions in engineering and industrial projects. One common resource-allocation methodology is the resource-assignment algorithm. This algorithm can be used to enhance the quality of resource-allocation decisions. Suppose there are n tasks that must be performed by n workers. The cost of worker i performing task j is c_{ij}. It is desirable to assign workers to tasks in a fashion that minimizes the cost of completing the tasks. This problem scenario is referred to as the assignment problem. The technique for finding the optimal solution to the problem is called the assignment method. The assignment method is an iterative procedure that arrives at the optimal solution by improving on a trial solution at each stage of the procedure.

The assignment method can be used to achieve an optimal assignment of resources to specific tasks in an industrial project. Although the assignment method is cost-based, task duration can be incorporated into the modeling in terms of time–cost relationships. The objective is to minimize the total cost of the project. Thus, the formulation of the assignment problem is as shown below:

Let
$x_{ij} = 1$ if worker i is assigned to task j, $j = 1, 2, ..., n$
$x_{ij} = 0$ if worker i is not assigned to task j
$c_{ij} = $ cost of worker i performing task j

Minimize: $$z = \sum_{i=1}^{n} \sum_{j=1}^{n} c_{ij} x_{ij}$$

Subject to: $$\sum_{j=1}^{n} x_{ij} = 1, \qquad i = 1,2,...,n$$

$$\sum_{i=1}^{n} x_{ij} = 1, \qquad j = 1,2,...,n$$

$$x_{ij} \geq 0, \quad i,j = 1,2,...,n$$

The preceding formulation uses the non-negativity constraint, $x_{ij} \geq 0$, instead of the integer constraint, $x_{ij} = 0$ or 1. However, the solution of the model will still be integer-valued. Hence, the assignment problem is a special case of the common transportation problem in operations research, with the number of sources (m) = number of targets (n), $S_i = 1$ (supplies), and $D_i = 1$ (demands). The basic requirements of an assignment problem are as follows:

1. There must be two or more tasks to be completed.
2. There must be two or more resources that can be assigned to the tasks.
3. The cost of using any of the resources to perform any of the tasks must be known.
4. Each resource is to be assigned to one and only one task.

If the number of tasks to be performed is greater than the number of workers available, we will need to add *dummy workers* to balance the problem. Similarly, if the number of workers is greater than the number of tasks, we will need to add *dummy tasks* to balance the problem. If there is no problem of overlapping, a worker's time may be split into segments so that the worker can be assigned more than one task. In this case, each segment of the worker's time will be modeled as a separate resource in the assignment problem. Thus, the assignment problem can be extended to consider partial allocation of resource units to multiple tasks.

The assignment model is solved by a method known as the *Hungarian method*, which is a simple iterative technique. Details of the assignment problem and its solution techniques can be found in operations-research texts. As an example, suppose five workers are to be assigned to five tasks on the basis of the cost matrix presented in Table 1.1. Task 3 is a machine-controlled task with a fixed cost of $800 regardless of the specific worker to whom it is assigned. Using the assignment method, we obtain the optimal solution presented in Table 1.2, which indicates the following:

$$x_{15} = 1, x_{23} = 1, x_{31} = 1, x_{44} = 1, \text{ and } x_{52} = 1$$

Thus, the minimum total cost (TC) is given by

$$\text{TC} = c_{15} + c_{23} + c_{31} + c_{44} + c_{52} = \$(400 + 800 + 300 + 400 + 350) = \$2,250$$

Table 1.1 Cost Matrix for Resource Assignment Problem

Worker	Task 1	Task 2	Task 3	Task 4	Task 5
1	300	200	800	500	400
2	500	700	800	1250	700
3	300	900	800	1000	600
4	400	300	800	400	400
5	700	350	800	700	900

Table 1.2 Solution to Resource Assignment Problem

Worker	Task 1	Task 2	Task 3	Task 4	Task 5
1	0	0	0	0	1
2	0	0	1	0	0
3	1	0	0	0	0
4	0	0	0	1	0
5	0	1	0	0	0

1.3   *Economics of resource utilization*

In industrial operations that are subject to risk and uncertainty, probability information can be used to analyze resource utilization characteristics of the operations. Suppose the level of availability of a resource is probabilistic in nature. For simplicity, we will assume that the level of availability, X, is a continuous variable whose probability density function is defined by $f(x)$. This is true for many resource types, ranging from funds and natural resources to raw materials. If we are interested in the probability that resource availability will be within a certain range of x_1 and x_2, then the required probability can be computed as follows:

$$P\left(x_1 \leq X \leq x_2\right) = \int_{x_1}^{x_2} f\left(x\right) dx$$

Similarly, a probability density function can be defined for the utilization level of a particular resource. If we denote the utilization level by U and its probability density function by $f(u)$, then we can calculate the probability that the utilization will exceed a certain level, u_0, by the following expression:

$$P\left(U \geq u_0\right) = \int_{u_0}^{\infty} f\left(u\right) du$$

Suppose that a critical resource is leased for a large project. There is a graduated cost associated with using the resource at a certain percentage level U. The cost is specified as $10,000 per 10% increment in utilization level above 40%. A flat cost of $5,000 is charged for utilization levels below 40%. The utilization intervals and the associated costs are as follows:

U < 40%, $5,000
40% ≤ U < 50%, $10,000
50% ≤ U < 60%, $20,000
60% ≤ U < 70%, $30,000
70% ≤ U < 80%, $40,000
80% ≤ U < 90%, $50,000
90% ≤ U < 100%, $60,000

Thus, a utilization level of 50% will cost $20,000, whereas a level of 49.5% will cost $10,000. Suppose the utilization level is a normally distributed random variable with a mean of 60% and a variance of 16% squared and that we are interested in finding the expected cost of using this resource. The solution procedure involves finding the probability that the utilization level will fall within each of the specified ranges. The expected value formula will then be used to compute the expected cost as shown below:

$$E[C] = \sum_k x_k P(x_k)$$

where x_k represents the kth interval of utilization. The standard deviation of utilization is 4%. Thus, we have the following:

$$P(U < 40) = P\left(z \le \frac{40 - 60}{4}\right) = P(z \le -5) = 0.0$$

$$P(40 \le U < 50) = 0.0062$$

$$P(50 \le U < 60) = 0.4938$$

$$P(60 \le U < 70) = 0.4938$$

$$P(70 \le U < 80) = 0.0062$$

$$P(80 \le U < 90) = 0.0$$

$$E(C) = \$5,000(0.0) + \$10,000(0.0062) + \$20,000(0.4938)$$

$$+ \$30,000(0.4938) + \$40,000(0.0062) + \$50,000(0.0)$$

$$= \$25,000$$

Based on these calculations, it can be expected that leasing this critical resource will cost $25,000 in the long run. A decision can be made as to whether to lease the resource, buy it, or substitute another resource for it, based on the information gained from this calculation.

1.4   Minimum annual revenue requirement (MARR) analysis

Companies evaluating capital expenditures for proposed projects must weigh the expected benefits against the initial and expected costs over the life cycle of the project. One method that is often used is MARR analysis. Using the information about costs, interest payments, recurring expenditures, and other project-related financial obligations, the minimum annual revenue required by a project can be evaluated. We can compute the break-even point of the project. The break-even point is then used to determine the level of revenue that must be produced by the project in order for it to be profitable. The analysis can be done with either the *flow-through* method or the *normalizing* method.

The factors to be included in MARR analysis are initial investment, book salvage value, tax salvage value, annual project costs, useful life for bookkeeping purposes, book depreciation method, tax depreciation method, useful life for tax purposes, rate of return on equity, rate of return on debt, capital interest rate, debt ratio, and investment tax credit. Computational details on these factors are presented in subsequent chapters of this book. This section presents an overall illustrative example of how companies use MARR as a part of their investment decisions.

The minimum annual revenue requirement for any year n may be determined by means of the net cash flows expected for that year:

$$\text{Net Cash Flow} = \text{Income} - \text{Taxes} - \text{Principal Amount Paid}$$

That is,

$$X_n = (G - C - I) - I - P$$

where

X_n = annual revenue for year n
G = gross income for year n
C = expenses for year n
I = interest payment for year n
t = taxes for year n
P = principal payment for year n

Rewriting the equation yields

$$G = X_n + C + I + t + P$$

The preceding equation assumes that there are no capital requirements, salvage value considerations, or working capital changes in year n. For the minimum annual gross income, the cash flow, X_n, must satisfy the following relationship:

$$X_n = D_e + f_n$$

where

D_e = recovered portion of the equity capital
f_n = return on the unrecovered equity capital

It is assumed that the total equity and debt capital recovered in a year are equal to the book depreciation, D_b, and that the principal payments are a constant percentage of the book depreciation. That is,

$$P = c(D_b)$$

where c is the debt ratio. The recovery of equity capital is, therefore, given by the following:

$$D_e = (1 - c)D_b$$

The annual returns on equity, f_n, and interest, I, are based on the unrecovered balance as follows:

$$f_n = (1 - c)k_e(BV_{n-1})$$

$$I = ck_d(BV_{n-1})$$

where

c = debt ratio
k_e = required rate of return on equity
k_d = required rate of return on debt capital
BV_{n-1} = book value at the beginning of year n

Based on the preceding equations, the minimum annual gross income, or revenue requirement, for year n can be represented as

$$R = D_b + f_n + C \mid I + t$$

An expression for taxes, t, is given by

$$t = (G - C - D_t - I) T$$

where

D_t = depreciation for tax purposes
T = tax rate

If the expression for R is substituted for G in the preceding equation, the following alternate expression for t can be obtained:

$$t = [T/(1 - T)](D_b + f_n - D_t)$$

The calculated minimum annual revenue requirement can be used to evaluate the economic feasibility of a project. An example of a decision criterion that may be used for that purpose is presented as follows:

Decision Criterion: If expected gross incomes are greater than MARR, then the project is considered to be economically acceptable, and the project investment is considered to be potentially profitable. Economic acceptance should be differentiated from technical acceptance, however. If, of course, other alternatives being considered have similar results, a comparison based on the margin of difference (i.e., incremental analysis) between the expected gross incomes and minimum annual requirements must be made. There are two extensions to the basic analysis procedure presented in the preceding text. They are the *flow-through method* and the *normalizing method*.

1.4.1   Flow-through method of MARR

This extension of the basic revenue requirement analysis allocates credits and costs in the year that they occur. That is, there are no deferred taxes, and the investment tax credit is not amortized. Capitalized interest is taken as an expense in the first year. The resulting equation for calculating the minimum annual revenue requirements is:

$$R = D_b + f_e + I + gP + C + t$$

where the required return on equity is given by the following:

$$f_e = k_e(1 - c)K_{n-1}$$

where

$\qquad k_e$ = implied cost of common stock
$\qquad c$ = debt ratio
$\qquad K_{n-1}$ = chargeable investment for the preceding year
$\qquad K_n = K_{n-1} - D_b$ (with K_0 = initial investment)
$\qquad g$ = capitalized interest rate

The capitalized interest rate is usually set by federal regulations. The debt interest is given by

$$I = (c)k_d K_{n-1}$$

where k_d = after-tax cost of capital.

The investment tax credit is calculated as follows:

$$C_t = i_t P$$

where *i* is the investment tax credit. Costs, *C*, are estimated totals that include such items as ad valorem taxes, insurance costs, operation costs, and maintenance costs. The taxes for the flow-through method are calculated as

$$t = \frac{T}{1-T}(f_e + D_b - D_i) - \frac{C}{1-T}$$

1.4.2 Normalizing method of MARR

The normalizing method differs from the flow-through method in that deferred taxes are utilized. These deferred taxes are sometimes included as expenses in the early years of the project and then as credits in later years. This *normalized* treatment of the deferred taxes is often used by public utilities to minimize the potential risk of changes in tax rules that may occur before the end of the project but are unforeseen at the start of the project. Also, the interest paid on the initial investment cost is capitalized. That is, it is taken as a tax deduction in the first year of the project and then amortized over the life of the project to spread out the interest costs. The resulting minimum annual revenue requirement is expressed as

$$R = D_b + d_t + C_t - A_t + I + f_e + t + C$$

where the depreciation schedules are based on the following capitalized investment cost:

$$K = P + gP$$

with *P* and *g* as previously defined. The deferred taxes, *d*, are the difference in taxes that result from using an accelerated depreciation model instead of a straight-line rate over the life of the project. That is,

$$d_t = (D_t - D_s)T$$

where

D_t = accelerated depreciation for tax purposes
D_s = straight-line depreciation for tax purposes

The amortized investment tax credit, A_t, is spread over the life of the project, *n*, and is calculated as follows:

$$A_t = \frac{C_t}{n}$$

The debt interest is similar to the earlier equation for capitalized interest. However, the chargeable investment differs by taking into account the investment tax credit, deferred taxes, and the amortized investment tax credit. The resulting expressions are:

$$I = k_d(c)K_{n-1}$$

$$K_n = K_{n-1} - D_b - C_t - d_t - A_t$$

In this case, the expression for taxes, t, is given by the following:

$$t = \frac{T}{1-T}\left(f_e + D_b + d_t + C_t - A_t - D_t - gP\right) - \frac{C_t}{1-T}$$

The differences between the procedures for calculating the minimum annual revenue requirements for the flow-through and the normalizing methods yield some interesting and important details. If the MARRs are converted to uniform annual amounts (leveled), a better comparison between the effects of the calculations for each method can be made. For example, the MARR data calculated by using each method are presented in Table 1.3.

The annual MARR values are denoted by R_n, and the uniform annual amounts are denoted by R_u. The uniform amounts are found by calculating the present value for each early amount and then converting that total amount to equal yearly amounts over the same span of time. For a given investment, the flow-through method will produce a smaller leveled minimum annual revenue requirement. This is because the normalized data include an amortized investment tax credit as well as deferred taxes. The yearly data for the flow-through method should give values closer to the actual cash flows because credits and costs are assigned in the year in which they occur and not up front, as in the normalizing method.

Table 1.3 Normalizing vs. Flow-through Revenue Analysis

Year	Normalizing		Flow-through	
	R_n	R_u	R_n	R_u
1	7135	5661	5384	5622
2	6433	5661	6089	5622
3	5840	5661	5913	5622
4	5297	5661	5739	5622
5	4812	5661	5565	5622
6	4380	5661	5390	5622
7	4005	5661	5214	5622
8	3685	5661	5040	5622

The normalizing method, however, provides for a faster recovery of the project investment. For this reason, this method is often used by public utility companies when establishing utility rates. The normalizing method also agrees better, in practice, with the required accounting procedures used by utility companies than does the flow-through method. Return on equity also differs between the two methods. For a given internal rate of return, the normalizing method will give a higher rate of return on equity than will the flow-through method. This difference occurs because of the inclusion of deferred taxes in the normalizing method.

Illustrative Example

Suppose we have the following data for a project. It is desired to perform a revenue requirement analysis using both the flow-through and the normalizing methods.

> Initial project cost = $100,000
> Book salvage value = $10,000
> Tax salvage value = $10,000
> Book depreciation model = Straight line
> Tax depreciation model = Sum-of-years digits
> Life for book purposes = 10 years
> Life for tax purposes = 10 years
> Total costs per year = $4,000
> Debt ratio = 40%
> Required return on equity = 20%
> Required return on debt = 10%
> Tax rate = 52%
> Capitalized interest = 0%
> Investment tax credit = 0%

Table 1.4, Table 1.5, Table 1.6, and Table 1.7 show the differences between the normalizing and flow-through methods for the same set of data. The different treatments of capital investment produced by the investment tax credit can be seen in the tables as well as in Figure 1.1, Figure 1.2, Figure 1.3, and Figure 1.4.

There is a big difference in the distribution of taxes because most of the taxes are paid early in the investment period with the normalizing method, but taxes are deferred with the flow-through method. The resulting MARR requirements are larger for the normalizing method early in the period. However, there is a more gradual decrease with the flow-through method. Therefore, the use of the flow-through method does not place as great a demand on the project to produce high revenues early in the project's life cycle as does the normalizing method. Also, the normalizing method produces a lower rate of return on equity. This fact may be of particular interest to shareholders.

Table 1.4 Part One of MARR Analysis

Year	Tax Depreciation		Deferred Taxes	
	Normalizing	Flow-through	Normalizing	Flow-through
1	16,363.64	16,363.64	3,829.09	None
2	14,727.27	14,727.27	2,978.18	
3	13,090.91	13,090.91	2,127.27	
4	11,454.55	11,454.55	1,276.36	
5	9,818.18	9,818.18	425.45	
6	8,181.82	8,181.82	425.45	
7	6,545.45	6,545.45	1,276.36	
8	4,909.09	4,909.09	2,127.27	
9	3,272.73	3,272.73	2,978.18	
10	1,636.36	1,636.36	3,829.09	

Table 1.5 Part Two of MARR Analysis

Year	Capitalized Investment		Taxes	
	Normalizing	Flow-through	Normalizing	Flow-through
	100,000.00	100,000.00	—	—
1	87,170.91	91,000.00	9,170.91	5,022.73
2	75,192.73	92,000.00	8,354.04	5,625.46
3	64,065.46	73,000.00	7,647.78	6,228.18
4	53,789.90	64,000.00	7,052.15	6,830.91
5	44,363.64	55,000.00	6,567.13	7,433.64
6	35,789.09	46,000.00	6,192.73	8,036.36
7	28,065.45	37,000.00	5,928.94	8,639.09
8	21,192.72	28,000.00	5,775.78	9,241.82
9	15,170.90	19,000.00	5,733.24	9,844.55
10	10,000.00	10,000.00	5,801.31	10,447.27

Table 1.6 Part Three of MARR Analysis

Year	Return on Debt		Return of Equity	
	Normalizing	Flow-through	Normalizing	Flow-through
1	4,000.00	4,000.00	12,000.00	12,000.00
2	3,486.84	3,640.00	10,460.51	10,920.00
3	3,007.71	3,280.00	9,023.13	9,840.00
4	2,562.62	2,920.00	7,687.86	8,760.00
5	2,151.56	2,560.00	6,454.69	7,680.00
6	1,774.55	2,200.00	5,323.64	6,600.00
7	1,431.56	1,840.00	4,294.69	5,520.00
8	1,122.62	1,480.00	3,367.85	4,440.00
9	847.71	1,120.00	2,543.13	3,360.00
10	606.84	760.00	1,820.51	2,280.00

Table 1.7 Part Four of MARR Analysis

	Minimum Annual Revenues	
Year	Normalizing	Flow-through
1	42,000.00	34,022.73
2	38,279.56	33,185.45
3	34,805.89	32,348.18
4	31,578.98	31,510.91
5	28,598.84	30,673.64
6	25,865.45	29,836.36
7	23,378.84	2,899.09
8	21,138.98	28,161.82
9	19,145.89	27,324.55
10	17,399.56	26,487.27

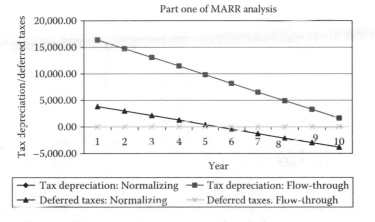

Figure 1.1 Plot of part one of MARR analysis.

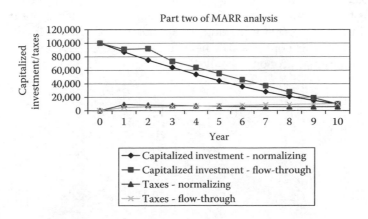

Figure 1.2 Plot of part two of MARR analysis.

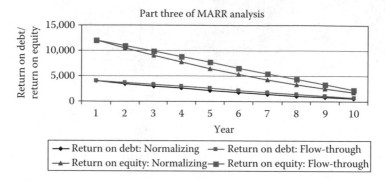

Figure 1.3 Plot of part three of MARR analysis.

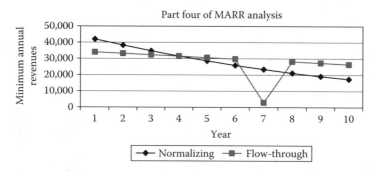

Figure 1.4 Plot of part four of MARR analysis.

This chapter has presented selected general techniques of applied economic analysis for industrial projects. Subsequent chapters present specific topics within the general body of knowledge for industrial economic analysis, focusing primarily on computational techniques. Applied economic analysis techniques, as presented in this chapter, are useful in engineering and industrial projects for making crucial business decisions involving buy, make, rent, or lease options. The next chapter covers cost concepts relevant for computational economic analysis.

chapter two

Cost concepts and techniques

The term *cost management* refers, in a project environment, to the functions required to maintain effective financial control of the project throughout its life cycle. There are several cost concepts that influence the economic aspects of managing engineering and industrial projects. Within a given scope of analysis, there may be a combination of different types of cost aspects to consider. These cost aspects include the ones defined here:

Actual cost of work performed: The cost actually incurred and recorded in accomplishing the work performed within a given period of time.

Applied direct cost: The amounts recognized in the time period associated with the consumption of labor, material, and other direct resources, without regard to the date of commitment or the date of payment. These amounts are to be charged to work-in-process (WIP) when resources are actually consumed, material resources are withdrawn from inventory for use, or material resources are received and scheduled for use within 60 d.

Budgeted cost for work performed: The sum of the budgets for completed work plus the appropriate portion of the budgets for level of effort and apportioned effort. Apportioned effort is that, which by itself is not readily divisible into short-span work packages, but is related in direct proportion to measured effort.

Budgeted cost for work scheduled: The sum of budgets for all work packages and planning packages scheduled to be accomplished (including work in process) plus the amount of level of effort and apportioned effort scheduled to be accomplished within a given period of time.

Direct cost: Cost that is directly associated with actual operations of a project. Typical sources of direct costs are direct material costs and direct labor costs. Direct costs are those that can be reasonably measured and allocated to a specific component of a project.

Economies of scale: A reduction of the relative weight of the fixed cost in total cost by increasing output quantity. This helps to reduce the final unit cost of a product. Economies of scale are often simply referred to as the savings due to *mass production.*

Estimated cost at completion: The actual direct costs, plus indirect costs that can be allocated to the contract, plus estimated costs (direct and indirect) for authorized work remaining.

First cost: The total initial investment required to initiate a project or the total initial cost of the equipment needed to start the project.

Fixed cost: A cost incurred irrespective of the level of operation of a project. Fixed costs do not vary in proportion to the quantity of output. Examples of costs that make up the fixed cost of a project are administrative expenses, certain types of taxes, insurance cost, depreciation cost, and debt-servicing cost. These costs usually do not vary in proportion to quantity of output.

Incremental cost: The additional cost of changing the production output from one level to another. Incremental costs are normally variable costs.

Indirect cost: A cost that is indirectly associated with project operations. Indirect costs are those that are difficult to assign to specific components of a project. An example of an indirect cost is the cost of computer hardware and software needed to manage project operations. Indirect costs are usually calculated as a percentage of a component of direct costs. For example, the indirect costs in an organization may be computed as 10% of direct labor costs.

Life-cycle cost: The sum of all costs, recurring and nonrecurring, associated with a project during its entire life cycle.

Maintenance cost: A cost that occurs intermittently or periodically and is used for the purpose of keeping project equipment in good operating condition.

Marginal cost: The additional cost of increasing production output by one additional unit. The marginal cost is equal to the slope of the total cost curve or line at the current operating level.

Operating cost: A recurring cost needed to keep a project in operation during its life cycle. Operating costs may consist of such items as labor cost, material cost, and energy cost.

Opportunity cost: The cost of forgoing the opportunity to invest in a venture that would have produced an economic advantage. Opportunity costs are usually incurred due to limited resources that make it impossible to take advantage of all investment opportunities. This is often defined as the cost of the best rejected opportunity. Opportunity costs can also be incurred due to a missed opportunity rather than due to an intentional rejection. In many cases, opportunity costs are hidden or implied because they typically relate to future events that cannot be accurately predicted.

Overhead cost: A cost incurred for activities performed in support of the operations of a project. The activities that generate overhead costs support the project efforts rather than contribute directly to the project goal. The handling of overhead costs varies widely from company to company. Typical overhead items are electric power cost, insurance premiums, cost of security, and inventory-carrying cost.

Standard cost: A cost that represents the normal or expected cost of a unit of the output of an operation. Standard costs are established in advance. They are developed as a composite of several component costs, such as direct labor cost per unit, material cost per unit, and allowable overhead charge per unit.

Sunk cost: A cost that occurred in the past and cannot be recovered under the present analysis. Sunk costs should have no bearing on the prevailing economic analysis and project decisions. Ignoring sunk costs is always a difficult task for analysts. For example, if $950,000 was spent 4 years ago to buy a piece of equipment for a technology-based project, a decision on whether or not to replace the equipment now should not consider that initial cost. However, uncompromising analysts might find it difficult to ignore so much money. Similarly, an individual making a decision on selling a personal automobile would typically try to relate the asking price to what was paid for the automobile when it was acquired. This is wrong under the strict concept of sunk costs.

Total cost: The sum of all the variable and fixed costs associated with a project.

Variable cost: A cost that varies in direct proportion to the level of operation or quantity of output. For example, the costs of material and labor required to make an item are classified as variable costs because they vary with changes in level of output.

2.1 Project cost estimation

Cost estimation and budgeting help establish a strategy for allocating resources in project planning and control. There are three major categories of cost estimation for budgeting based on the desired level of accuracy: *order-of-magnitude estimates*, *preliminary cost estimates*, and *detailed cost estimates*. Order-of-magnitude cost estimates are usually gross estimates based on the experience and judgment of the estimator. They are sometimes called "ballpark" figures. These estimates are typically made without a formal evaluation of the details involved in the project. Order-of-magnitude estimates can range, in terms of accuracy, from 50 to +50% of the actual cost. These estimates provide a quick way of getting cost information during the initial stages of a project.

50% (Actual Cost) ≤ Order-of-Magnitude Estimate ≤ 150% (Actual Cost)

Preliminary cost estimates are also gross estimates but with a higher level of accuracy. In developing preliminary cost estimates, more attention is paid to some selected details of the project. An example of a preliminary cost estimate is the estimation of expected labor cost. Preliminary estimates are useful for evaluating project alternatives before final commitments are made. The level of accuracy associated with preliminary estimates can range from 20 to +20% of the actual cost.

80% (Actual Cost) ≤ Preliminary Estimate ≤ 120% (Actual Cost)

Detailed cost estimates are developed after careful consideration is given to all the major details of a project. Considerable time is typically needed to obtain detailed cost estimates. Because of the amount of time and effort needed to develop detailed cost estimates, the estimates are usually developed after there is firm commitment that the project will happen. Detailed cost estimates are also important for evaluating actual cost performance during the project. The level of accuracy associated with detailed estimates normally ranges from 5 to +5% of the actual cost.

95% (Actual Cost) ≤ Detailed Cost ≤ 105% (Actual Cost)

There are two basic approaches to generating cost estimates. The first one is a variant approach, in which cost estimates are based on variations of previous cost records. The other approach is the generative cost estimation, in which cost estimates are developed from scratch without taking previous cost records into consideration.

2.1.1   Optimistic and pessimistic cost estimates

Using an adaptation of the PERT formula, we can combine optimistic and pessimistic cost estimates. Let

O = optimistic cost estimate
M = most likely cost estimate
P = pessimistic cost estimate

Then, the estimated cost can be estimated as

$$E[C] = \frac{O + 4M + P}{6}$$

The cost variance can be estimated as

$$V[C] = \left[\frac{P - O}{6}\right]^2$$

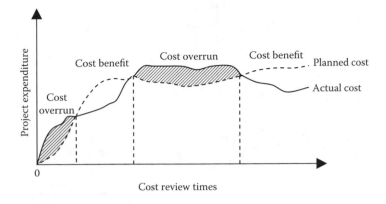

Figure 2.1 Evaluation of actual and projected cost.

2.2 Cost monitoring

As a project progresses, costs can be monitored and evaluated to identify areas of unacceptable cost performance. Figure 2.1 shows a plot of cost vs. time for projected cost and actual cost. The plot permits quick identification when cost overruns occur in a project. Plots similar to those presented in the figure may be used to evaluate the cost, schedule, and time performances of a project. An approach similar to the profit ratio presented earlier may be used together with the plot to evaluate the overall cost performance of a project over a specified planning horizon. The following is a formula for the *cost performance index* (CPI):

$$CPI = \frac{\text{Area of cost benefit}}{\text{Area of cost benefit} + \text{area of cost overrun}}$$

As in the case of the profit ratio, CPI may be used to evaluate the relative performances of several project alternatives or to evaluate the feasibility and acceptability of an individual alternative. In Figure 2.2, we present another cost monitoring tool: the cost control pie chart. This type of chart is used to track the percentage of cost going into a specific component of a project. Control limits can be included in the pie chart to identify out-of-control cost situations. The example in Figure 2.2 shows that 10% of total cost is tied up in supplies. The control limit is located at 12% of total cost. Hence, the supplies expenditure is within control (so far, at least).

2.3 Project balance technique

One other approach to monitoring cost performance is the project balance technique, one that helps in assessing the economic state of a project at a

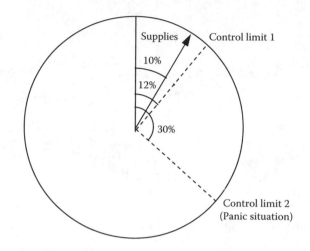

Figure 2.2 Cost control pie chart.

desired point in time in the life cycle of the project. It calculates the net cash flow of a project up to a given point in time. The project balance is calculated as follows:

$$B(i)_t = S_t - P(1+i)^t + \sum_{k=1}^{t} PW_{income}(i)_k$$

where

$B(i)_t$ = project balance at time t at an interest rate of $i\%$ per period

PW income $(i)_t$ = present worth of net income from the project up to time t

P = initial cost of the project

S_t = salvage value at time t

The project balance at time t gives the net loss or net profit associated with the project up to that time.

2.4 Cost and schedule control systems criteria

Contract management involves the process by which goods and services are acquired, utilized, monitored, and controlled in a project. Contract management addresses the contractual relationships from the initiation of a project to its completion (i.e., completion of services and/or hand-over of deliverables). Some of the important aspects of contract management include

- Principles of contract law
- Bidding process and evaluation

- Contract and procurement strategies
- Selection of source and contractors
- Negotiation
- Worker safety considerations
- Product liability
- Uncertainty and risk management
- Conflict resolution

In 1967, the U.S. Department of Defense (DOD) introduced a set of 35 standards or criteria with which contractors must comply under cost or incentive contracts. The system of criteria is referred to as the *Cost and Schedule Control Systems Criteria* (C/SCSC). Many government agencies now require compliance with C/SCSC for major contracts. The system presents an integrated approach to cost and schedule management, and its purpose is to manage the government's risk of cost overruns. Now widely recognized and used in major project environments, it is intended to facilitate greater uniformity and provide advance warning about impending schedule or cost overruns.

The topics addressed by C/SCSC include cost estimating and forecasting, budgeting, cost control, cost reporting, earned value analysis, resource allocation and management, and schedule adjustments. The important link between all of these is the dynamism of the relationship between performance, time, and cost. Such a relationship is represented in Figure 2.3. This is essentially a multiobjective problem. Because performance, time, and cost objectives cannot be satisfied equally well, concessions or compromises need to be worked out in implementing C/SCSC.

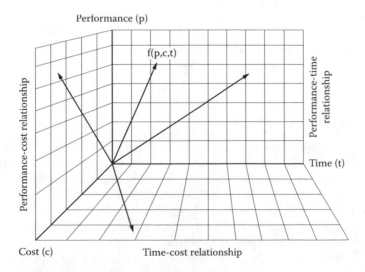

Figure 2.3 Performance–cost–time relationships for C/SCSC.

Another dimension of the performance–time–cost relationship is represented by the U.S. Air Force's R&M 2000 Standard, which addresses the reliability and maintainability of systems. R&M 2000 is intended to integrate reliability and maintainability into the performance, cost, and schedule management for government contracts. C/SCSC and R&M 2000 together constitute an effective guide for project design.

To comply with C/SCSC, contractors must use standardized planning and control methods that are based on *earned value*. This refers to the actual dollar value of work performed at a given point in time, compared to the planned cost for the work. It is different from the conventional approach of measuring actual vs. planned, which is explicitly forbidden by C/SCSC. In the conventional approach, it is possible to misrepresent the actual content (or value) of the work accomplished. The work rate analysis technique presented in this book can be useful in overcoming the deficiencies of the conventional approach. C/SCSC is developed on a work content basis, using the following factors:

- The actual cost of work performed (ACWP), which is determined on the basis of the data from the cost accounting and information systems
- The budgeted cost of work scheduled (BCWS) or baseline cost determined by the costs of scheduled accomplishments
- The budgeted cost of work performed (BCWP) or earned value, the actual work of effort completed as of a specific point in time

The following equations can be used to calculate cost and schedule variances for a work package at any point in time.

Cost variance = BCWP − ACWP
Percent cost variance = (Cost variance/BCWP) · 100
Schedule variance = BCWP − BCWS
Percent schedule variance = (Schedule variance/BCWS) · 100
ACWP and remaining funds = Target cost (TC)
ACWP + cost to complete = Estimated cost at completion (EAC)

2.5   Sources of capital

Financing a project means raising capital for the project. "Capital" is a resource consisting of funds available to execute a project, and it includes not only privately owned production facilities but also public investment. Public investments provide the infrastructure of the economy, such as roads, bridges, water supply, and so on. Other public capital that indirectly supports production and private enterprise includes schools, police stations, a central financial institution, and postal facilities.

If the physical infrastructure of the economy is lacking, the incentive for private entrepreneurs to invest in production facilities is likely to be lacking also. Government and/or community leaders can create the atmosphere for free enterprise by constructing better roads, providing better public safety and better facilities, and by encouraging ventures that will assure adequate support services.

As far as project investment is concerned, what can be achieved with project capital is very important. The avenues for raising capital funds include banks, government loans or grants, business partners, cash reserves, and other financial institutions. The key to the success of the free-enterprise system is the availability of capital funds and the availability of sources to invest the funds in ventures that yield products needed by the society. Some specific ways that funds can be made available for business investments are discussed in the following text.

2.6   Commercial loans

Commercial loans are the most common sources of project capital. Banks should be encouraged to lend money to entrepreneurs, particularly those who are just starting new businesses. Government guarantees may be provided to make it easier for an enterprise to obtain the needed funds.

2.7   Bonds and stocks

Bonds and stocks are also common sources of capital. National policies regarding the issuance of bonds and stocks can be developed to target specific project types in order to encourage entrepreneurs.

2.8   Interpersonal loans

Interpersonal loans are an unofficial means of raising capital. In some cases, there may be individuals with enough personal funds to provide personal loans to aspiring entrepreneurs. But presently, there is no official mechanism that handles the supervision of interpersonal business loans. If a supervisory body existed at a national level, wealthy citizens might be less apprehensive about lending money to friends and relatives for business purposes. Individual wealthy citizens could, thus, become a strong source of business capital. *Venture capitalists* often operate as individuals or groups of individuals providing financing for entrepreneurial activities.

2.9   Foreign investment

Foreign investment can be attracted for local enterprises through government incentives, which may take such forms as attractive zoning permits, foreign exchange permits, or tax breaks.

2.10   Investment banks

The operations of investment banks are often established to raise capital for specific projects. Investment banks buy securities from enterprises and resell them to other investors. Proceeds from these investments may serve as a source of business capital.

2.11   Mutual funds

Mutual funds represent collective funds from a group of individuals. Such collective funds are often large enough to provide capital for business investments. Mutual funds may be established by individuals or under the sponsorship of a government agency. Encouragement and support should be provided for the group to spend the money for business investment purposes.

2.12   Supporting resources

The government may establish a clearinghouse of potential goods and services that a new project can provide. New entrepreneurs interested in providing these goods and services should be encouraged to start relevant enterprises and given access to technical, financial, and information resources to facilitate starting production operations. A good example of this is "partnership" financing whereby cooperating entities come together to fund capital-intensive projects. The case study in Chapter 13 illustrates an example of federal, state, and commercial bank partnership to finance a large construction project.

2.13   Activity-based costing

Activity-based costing (ABC) has emerged as an appealing costing technique in industry. The major motivation for adopting ABC is that it offers an improved method to achieve enhancements in operational and strategic decisions. ABC offers a mechanism to allocate costs in direct proportion to the activities that are actually performed. This is an improvement over the traditional way of generically allocating costs to departments. It also improves the conventional approaches to allocating overhead costs. The use of PERT/CPM, precedence diagramming, and critical resource diagramming can facilitate task decomposition to provide information for ABC. Some of the potential impacts of ABC on a production line include the following:

- Identification and removal of unnecessary costs
- Identification of the cost impact of adding specific attributes to a product
- Indication of the incremental cost of improved quality
- Identification of the value-added points in a production process
- Inclusion of specific inventory-carrying costs

- Provision of a basis for comparing production alternatives
- The ability to assess "what-if" scenarios for specific tasks

ABC is just one component of the overall activity-based management in an organization. Activity-based management involves a more global management approach to planning and control of organizational endeavors. This requires consideration for product planning, resource allocation, productivity management, quality control, training, line balancing, value analysis, and a host of other organizational responsibilities. Thus, although activity-based costing is important, one must not lose sight of the universality of the environment in which it is expected to operate. Frankly, there are some processes whose functions are so intermingled that separating them into specific activities may be difficult. Major considerations in the implementation of ABC include these:

- Resources committed to developing activity-based information and cost
- Duration and level of effort needed to achieve ABC objectives
- Level of cost accuracy that can be achieved by ABC
- Ability to track activities based on ABC requirements
- Handling the volume of detailed information provided by ABC
- Sensitivity of the ABC system to changes in activity configuration

Income analysis can be enhanced by the ABC approach as shown in Table 2.1. Similarly, instead of allocating manufacturing overhead on the basis of direct labor costs, an activity-based costing analysis can be done, as illustrated in the example presented in Table 2.2. Table 2.3 shows a more comprehensive use of ABC to compare product lines. The specific ABC cost components shown in Table 2.3 can be further broken down if needed. A spreadsheet analysis would indicate the impact on net profit as specific cost elements are manipulated. Based on this analysis, it is seen that Product Line A is the most profitable. Product Line B comes in second even though it has the highest total line cost. Figure 2.4 presents a graphical comparison of the ABC cost elements for the product lines.

2.14   Cost, time, and productivity formulas

This section presents a collection of common formulas useful for cost, time, and productivity analysis in manufacturing projects.

> *Average time to perform a task:* Based on learning-curve analysis, the average time required to perform a repetitive task is given by the following:

$$t_n = an^{-b}$$

Table 2.1 Sample of Project Income Statement

Statement of Income		
(In Thousands of Dollars, Except Per-Share Amounts)		
Two years ended December 31	2006	2007
Net sales	1,918,265	1,515,861
Costs and expenses		
Cost of sales	$1,057,849	$878,571
Research and development	72,511	71,121
Marketing and distribution	470,573	392,851
General and administrative	110,062	81,825
	1,710,995	1,424,268
Operating income	207,270	91,493
Consolidation of operations	(36,981)	
Interest and other income, net	9,771	17,722
Income before taxes	180,060	109,215
Provision for income taxes	58,807	45,115
Net income	$121,253	$64,100
Equivalent shares	61,880	60,872
Earnings per common share	$1.96	$1.05

where

t_n = cumulative average time resulting from performing the task n times

t_1 = time required to perform the task the first time

k = learning factor for the task (usually known or assumed)

The parameter k is a positive real constant whose magnitude is a function of the type of task being performed. A large value of k would cause the overall average time to drop quickly after just a few repetitions of the task. Thus, simple tasks tend to have large learning factors. Complex tasks tend to have smaller learning factors, thereby requiring several repetitions before significant reduction in time can be achieved.

Calculating the learning factor: If the learning factor is not known, it may be estimated from time observations by the following formula:

$$k = \frac{\log t_1 - \log t_n}{\log n}$$

Calculating total time: Total time, T_n, to complete a task n times, if the learning factor and the initial time are known, is obtained by multiplying the average time by the number of times. Thus,

$$T_n = t_1 n^{(1-k)}$$

Table 2.2 Activity-Based Cost Details for Industrial Project

	Unit Cost ($)	Cost Basis	Days Worked	Cost ($)
Labor				
Design engineer	200	Day	34	$6,800
Carpenter	150	Day	27	4,050
Plumber	175	Day	2	350
Electrician	175	Day	82	14,350
IS engineer	200	Day	81	16,200
Labor subtotal				41,750
Contractor				
Air conditioning	10,000	Fixed	5	10,000
Access flooring	5,000	Fixed	5	5,000
Fire suppression	7,000	Fixed	5	7,000
AT&T	1,000	Fixed	50	1,000
DEC reinstall	4,000	Fixed	2	4,000
DEC install	8,000	Fixed	7	8,000
VAX mover	1,100	Fixed	7	1,100
Transformer mover	300	Fixed	7	300
Contractor subtotal				36,400
Materials				
Site preparation	2,500	Fixed	—	2,500
Hardware	31,900	Fixed	—	31,900
Software	42,290	Fixed	—	42,290
Other	10,860	Fixed	—	10,860
Materials subtotal				87,550
Grand Total				165,700

Determining time for nth performance of a task: The time required to perform a task the *n*th time if given by

$$x_n = t_1(1-k)n^{-k}$$

Determining limit of learning effect: The limit of learning effect indicates the number of times of performance of a task at which no further improvement is achieved. This is often called the improvement ratio and is represented as

$$n \geq \frac{1}{1-r^{1/k}}$$

Determining improvement target: It is sometimes desired to achieve a certain level of improvement after so many performances of a task, given a certain learning factor, *k*. Supposing that it takes so many trials, n_1, to achieve a certain average time performance, y_1, and that it is

Table 2.3 ABC Comparison of Product Lines

ABC Cost Components	Product A	Product B	Product C	Product D
Direct labor	27,000.00	37,000.00	12,500.00	16,000.00
Direct materials	37,250.00	52,600.00	31,000.00	35,000.00
Supplies	1,500.00	1,300.00	3,200.00	2,500.00
Engineering	7,200.00	8,100.00	18,500.00	17,250.00
Material handling	4,000.00	4,200.00	5,000.00	5,200.00
Quality assurance	5,200.00	6,000.00	9,800.00	8,300.00
Inventory cost	13,300.00	17,500.00	10,250.00	11,200.00
Marketing	3,000.00	2,700.00	4,000.00	4,300.00
Equipment depreciation	2,700.00	3,900.00	6,100.00	6,750.00
Utilities	950.00	700.00	2,300.00	2,800.00
Taxes and insurance	3,500.00	4,500.00	2,700.00	3,000.00
Total line cost	**105,600.00**	**138,500.00**	**105,350.00**	**112,300.00**
Annual production	13,000.00	18,000.00	7,500.00	8,500.00
Cost/unit	8.12	7.69	14.05	13.21
Price/unit	9.25	8.15	13.25	11.59
Net profit/unit	1.13	0.46	0.80	1.62
Total line revenue	**120,250.00**	**146,700.00**	**99,375.00**	**98,515.00**
Net line profit	**14,650.00**	**8,200.00**	**5,975.00**	**13,785.00**

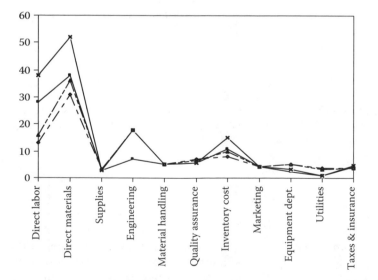

Figure 2.4 Activity-based comparison of product lines.

desired to calculate how many trials, n_2, would be needed to achieve a given average time performance, y_2, the following formula would be used:

$$n_2 = n_1 y_1^{1/k} y_2^{-1/k} = n_1 \left(y_2/y_1\right)^{-1/k} = n_1 r^{-1/k}$$

where the parameter, r, is referred to as the time improvement factor.

Calculation of number of machines needed to meet output: The number of machines needed to achieve a specified total output is calculated from the following formula:

$$N = \frac{1.67t\left(O_T\right)}{uH}$$

where
    N = number of machines
    t = processing time per unit (in minutes)
    O_T = total output per shift
    u = machine utilization ratio (in decimals)
    H = hours worked per day (8 times number of shifts)

Calculation of machine utilization: Machine idle times adversely affect utilization. The fraction of the time a machine is productively engaged is referred to as the utilization ratio and is calculated as follows:

$$u = \frac{h_a}{h_m}$$

where
    h_a = actual hours worked
    h_m = maximum hours a machine could work

The *percent utilization* is obtained as 100% times u.

Calculation of output to allow for defects: To allow for a certain fraction of defects in total output, use the following formula to calculate starting output:

$$Y = \frac{X}{1-f}$$

where
    Y = starting output
    X = target output
    f = fraction defective

Calculation of machine availability: The percent of time that a machine is available for productive work is calculated as follows:

$$A = \frac{o - u}{o}\left(100\%\right)$$

Practice problems for cost concepts and techniques

2.1 If 70 standard labor hours were required in a manufacturing plant to construct some equipment for the first time, and experience with similar products indicates a learning curve of 90%, how many hours are required for the 150th unit? For the 350th unit?

2.2 The standard screening minutes per patient in a hospital based on a time study of the 42nd patient is 18.7. If the learning curve based on previous experience of patients with a similar illness is 88%, (a) What was the number of minutes required for the first patient? (b) What is the estimated number of minutes needed for the 100th patient?

2.3 If you were considering implementing ABC in an oil-and-gas or manufacturing or service facility, what are some examples of data or information that you would require before full implementation?

2.4 If the data requirements in Problem 2.3 were very costly to obtain, could you consider partially implementing ABC? How would you decide which overhead factors to allocate with ABC?

2.5 What are some advantages and disadvantages of using the cost performance index (CPI) for monitoring project costs? Develop a discounted CPI formula for monitoring project costs.

chapter three

Fundamentals of economic analysis

Capital, in the form of money, is one of the factors that sustains business projects or ventures in the enterprise of producing wealth. However, it is necessary to intelligently consider the implications of committing capital to a business over a period of time; the discipline of economic analysis helps us achieve that aim. The time value of money is an important factor in economic consideration of projects. This is particularly crucial for long-term projects that are subject to changes in several cost parameters. Both the timing and quantity of cash flow are important for project management. The evaluation of a project alternative requires consideration of the initial investment, depreciation, taxes, inflation, economic life of the project, salvage value, and cash flow. Capital can be classified into two categories: equity and debt. Equity capital is owned by individuals and invested with the hope of making profit, whereas debt capital is borrowed from lenders such as banks. In this chapter, we explain the nature of capital, interest, and the fundamental concepts underlying the relationship between capital investments and the terms of those investments. These fundamental concepts play a central role throughout the rest of this book.

3.1   The economic analysis process

The process of economic analysis has the following basic components. The specific components will differ depending on the nature and prevailing circumstances of a project. This list is, however, representative of what an analyst might expect to encounter:

1. Problem identification
2. Problem definition
3. Development of metrics and parameters
4. Search for alternate solutions
5. Selection of the preferred solutions
6. Implementation of the selected solution
7. Monitoring and sustaining the project

3.2 Simple and compound interest rates

Interest rates are used to quantify the time value of money, which may be defined as the value of capital committed to a project or business over a period of time. Interest *paid* is the cost on borrowed money, and interest *earned* is the benefit on saved or invested money. Interest rates can be calculated as simple rates or compound rates. *Simple interest* is interest paid only on the principal, whereas *compound interest* is interest paid on both the principal and the accrued interest.

     Let

F_n = future value after n periods
P = initial investment amount (the principal)
I = interest rate per interest paid
n = number of investment or loan periods
r = nominal interest rate per year
m = number of compounding periods per year
i_a = effective interest rate per compounding period
i = effective interest rate per year

The expressions for computing interest amounts based on simple interest and compound interest are as follows:

$$\text{Simple interest:} \qquad I_n = P(i)(n)$$

$$\text{Compound interest:} \qquad I_n = iF_{(n-1)}$$

$F_{(n-1)}$ is the future value at period $(n - 1)$, which is the period immediately preceding the one for which the interest amount is being computed. The future values at time n are computed as:

$$\text{Simple interest:} \qquad F_n = P(1 + ni)$$

$$\text{Compound interest:} \qquad F_n = P(1 + i)^n$$

F_n is the accumulated value after n periods. At $n = 0$ and 1, the future values for both simple interest and compound interest are equal as shown in Table 3.1. In other words, simple and compound interest calculations yield the same results for F_n only when $n = 0$ or $n = 1$. Computationally,

$$F_n = P(1+in) \equiv F_n = P(1+i)^n$$

$$\Rightarrow P(1+in) = P(1+i)^n$$

$$\Rightarrow (1+in) = (1+i)^n$$

$$\Rightarrow (1+i)^n - (1+in) = 0$$

$$\Rightarrow n = 0 \ \ or \ \ n = 1$$

Table 3.1 Comparison of Simple Interest and Compound Interest Computations

N (periods)	Cash Flow	Simple Interest	Compound Interest
0		$F_0 = P + I_0$	$F_0 = P + I_0$
		$= P + P(i)(0)$	$= P + P(i)(0)$
		$= P$	$= P$
1		$F_1 = P + I_1$	$F_1 = F_0 + I_1$
		$= P + P(i)(1)$	$= F_0 + iF_0$
		$= P(1+i)$	$= F_0(1+i)$
			$= P(1+i)$
2		$F_2 = P + I_2$	$F_2 = F_1 + I_2$
		$= P + P(i)(2)$	$= F_1 + iF_1$
		$= P(1+2i)$	$F_1(1+i)$
			$P(1+i)(1+i)$
			$P(1+i)^2$
n		$F_n = P(1+ni)$	$F_n = P(1+i)^n$

It can be seen that compound interest calculations represent a compound sum of a series of one-period simple interest calculations. Because simple interest is not widely used in economic analysis, it will not be further discussed, other than to point out differences between it and compound interest.

3.3   Investment life for multiple returns

A topic that is often of intense interest in many investment scenarios is how long it will take a given amount to reach a certain multiple of its initial level. The "Rule of 72" is one simple approach to calculating how long it will take an investment to double in value at a given interest rate per period. The Rule of 72 gives the following formula for estimating the doubling period:

$$n = \frac{72}{i}$$

where i is the interest rate expressed in percentage. Referring to the single-payment compound amount factor, we can set the future amount equal to twice the present amount and then solve for n, the number of periods. That is, $F = 2P$. Thus,

$$2P = P(1+i)^n$$

Solving for n in the preceding equation yields an expression for calculating the exact number of periods required to double P:

$$n = \frac{\ln(2)}{\ln(1+i)}$$

where i is the interest rate expressed in decimals. When exact computation is desired, the length of time it would take to accumulate m multiple of P is expressed in its general form as:

$$n = \frac{\ln(m)}{\ln(1+i)}$$

where m is the desired multiple. For example, at an interest rate of 5% per year, the time it would take an amount P to double in value ($m = 2$) is 14.21 years. This, of course, assumes that the interest rate will remain constant throughout the planning horizon. Table 3.2 presents a tabulation of the values calculated from both approaches. Figure 3.1 shows a graphical comparison of the results from use of the Rule of 72 to use of the exact calculation.

3.4   Nominal and effective interest rates

The compound interest rate, which we will refer to as simply "interest rate," is used in economic analysis to account for the time value of money. Interest rates are usually expressed as a percentage, and the interest period (the time

Table 3.2 Evaluation of the Rule of 72

i%	n (Rule of 72)	n (Exact value)
0.25	288.00	277.61
0.50	144.00	138.98
1.00	72.00	69.66
2.00	36.00	35.00
5.00	14.20	17.67
8.00	9.00	9.01
10.00	7.20	7.27
12.00	6.00	6.12
15.00	4.80	4.96
18.00	4.00	4.19
20.00	3.60	3.80
25.00	2.88	3.12
30.00	2.40	2.64

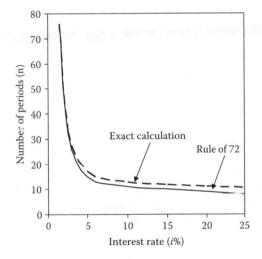

Figure 3.1 Evaluation of investment life for double return.

unit of the rate) is usually a year. However, interest rates can also be computed more than once a year. Compound interest rates can be quoted as *nominal interest rates* or as *effective interest rates*.

A *nominal interest rate* is the interest rate as quoted without considering the effect of any compounding. It is not the real interest rate used for economic analysis; however, it is usually the quoted interest rate because it is numerically smaller than the effective interest rate. It is equivalent to the *annual percentage rate* (APR), which is usually quoted for loan and credit-card purposes. The expression for calculating the nominal interest rate is as follows:

$$r = \left(\text{interest rate per period}\right) \times \left(\text{number of periods}\right)$$

The format for expressing r is as follows:

$$r\% \text{ per time period } t$$

The effective interest rate can be expressed either per year or per compounding period. It is the effective interest rate per year that is used in engineering economic analysis calculations. It is the annual interest rate taking into consideration the effect of any compounding during the year. It accounts for both the nominal rate and the compounding frequency. Effective interest rate *per year* is given by:

$$i = \left(1+i\right)^m - 1$$

$$= \left(1 + \frac{r}{m}\right)^m - 1$$

$$= \left(F/P, r/m\%, m\right) - 1$$

Effective interest rate *per compounding period* is given by:

$$i_a = \left(1+i\right)^{\frac{1}{m}} - 1 = \frac{r}{m}$$

When compounding occurs more frequently, the compounding period becomes shorter; hence, we have the phenomenon of continuous compounding. This situation can be seen in the stock markets. The effective interest rate for *continuous compounding* is given by:

$$i = e^r - 1$$

Note that the time period for i and r must be the same in using the preceding equations.

Example 3.1

The nominal annual interest rate of an investment is 9%. What is the effective annual interest rate if the interest is

1. Payable, or compounded, quarterly?
2. Payable, or compounded, continuously?

Solution

1. Using the effective interest rate formula, the effective annual interest rate compounded quarterly =

$$\left(1+\frac{0.09}{4}\right)^4 - 1 = 9.31\%$$

2. Using the equation for continuous rate, the effective annual interest rate compounded continuously = $e^{0.09} - 1 = 9.42\%$

The slight difference between each of these values and the nominal interest rate of 9% becomes a big concern if the period of computation is in the double digits. The effective interest rate must always be used in all computations. Therefore, a correct identification of the nominal and effective interest rates is very important. See the following example.

Example 3.2

Identify the following interest rate statements as either nominal or effective:

1. 14% per year
2. 1% per month, compounded weekly
3. Effective 15% per year, compounded monthly
4. 1.5% per month, compounded monthly
5. 20% per year, compounded semiannually

Solution

1. This is an *effective interest rate*. This may also be written as 14% per year, compounded yearly.
2. This is a *nominal interest rate* because the rate of compounding is not equal to the rate of interest time period.
3. This is an *effective interest for yearly rate*.
4. This is an *effective interest for monthly rate*. A new rate should be computed for yearly computations. This may also be written as 1.5% per month.
5. This is a *nominal interest rate* because the rate of compounding and the rate of interest time period are not the same.

3.5 *Cash-flow patterns and equivalence*

The basic reason for performing economic analysis is to provide information that helps in making choices between mutually exclusive projects competing for limited resources. The cost performance of each project will depend on the timing and levels of its expenditures. By using various techniques of computing cash-flow equivalence, we can reduce competing project cash flows to a common basis for comparison. The common basis depends, however, on the prevailing interest rate. Two cash flows that are equivalent at a given interest rate are not equivalent at a different interest rate. The basic techniques for converting cash flows from an interest rate at one point in time to the interest rate at another are presented in this section.

A *cash-flow diagram* (CFD) is a graphical representation of revenues (cash inflows) and expenses (cash outflows). If several cash flows occur during the same time period, a net cash-flow diagram is used to represent the differences in cash flows. Cash-flow diagrams are based on several assumptions:

- Interest rate is computed once in a time period.
- All cash flows occur at the end of the time period.
- All periods are of the same length.
- The interest rate and the number of periods are of the same length.
- Negative cash flows are drawn downward from the time line.
- Positive cash flows are drawn upward from the time line.

Cash-flow conversion involves the transfer of project funds from one point in time to another. There are several factors used in the conversion of cash flows.

Let:

P = cash flow value at the present time period. This usually occurs at time 0.

F = cash flow value at some time in the future.

A = a series of equal, consecutive, and end-of-period cash flow. This is also called annuity.

G = a uniform arithmetic gradient increase in period-by-period cash flow.

t = a measure of time period. It can be stated in years, months, or days.

n = the total number of time periods, which can be in days, weeks, months, or years.

i = interest rate time period expressed as a percentage.

In many cases, the interest rate used in performing economic analysis is set equal to the minimum attractive rate of return (MARR) of the decision maker. MARR is also sometimes referred to as the *hurdle rate*, the *required internal rate of return* (IRR), the *return on investment* (ROI), or the *discount rate*. The value of MARR is chosen with the objective of maximizing the economic performance of a project.

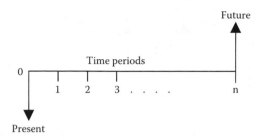

Figure 3.2 Single-payment compound amount cash flow.

3.6 Compound amount factor

The procedure for the single-payment compound amount factor finds a future sum of money, F, that is equivalent to a present sum of money, P, at a specified interest rate, i, after n periods. This is calculated as:

$$F = P(1 + i)^n$$

A graphical representation of the relationship between P and F is shown in Figure 3.2.

Example 3.3

A sum of $5,000 is deposited in a project account and is left there to earn interest for 15 years. If the interest rate per year is 12%, the compound amount after 15 years can be calculated as follows:

$$F = \$5,000(1 + 0.12)^{15}$$

$$= \$27,367.85$$

3.7 Present worth factor

The present worth factor computes P when F is given. It is obtained by solving for P in the equation for the compound amount factor. That is,

$$P = F(1 + i)^n$$

Suppose it is estimated that $15,000 would be needed to complete the implementation of a project five years in the future; how much should be deposited in a special project fund now so that the fund would accrue to the required $15,000 exactly in five years? If the special project fund pays interest at 9.2% per year, the required deposit would be:

$$P = \$15,000(1 + 0.092)^5$$

$$= \$9,660.03$$

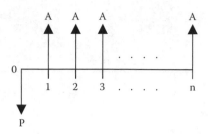

Figure 3.3 Uniform series cash flow.

3.8 *Uniform series present worth factor*

The uniform series present worth factor is used to calculate the present worth equivalent, P, of a series of equal end-of-period amounts, A. Figure 3.3 shows the uniform series cash flow. The derivation of the formula uses the finite sum of the present worths of the individual amounts in the uniform series cash flow, as follows. Some formulas for series and summation operations are presented in the Appendixes at the end of the book.

$$P = \sum_{t=1}^{n} A(1+i)^{-t}$$

$$= A\left[\frac{(1+i)^n - 1}{i(1+i)^n}\right]$$

Example 3.4

Suppose that the sum of $12,000 must be withdrawn from an account to meet the annual operating expenses of a multiyear project. The project account pays interest at 7.5% per year compounded on an annual basis. If the project is expected to last 10 years, how much must be deposited in the project account now so that the operating expenses of $12,000 can be withdrawn at the end of every year for 10 years? The project fund is expected to be depleted to zero by the end of the last year of the project. The first withdrawal will be made 1 year after the project account is opened, and no additional deposits will be made in the account during the project life cycle. The required deposit is calculated to be:

$$P = \$12,000\left[\frac{(1+0.075)^{10} - 1}{0.075(1+0.075)^{10}}\right]$$

$$= \$82,368.92$$

3.9　Uniform series capital recovery factor

The capital recovery formula is used to calculate the uniform series of equal end-of-period payments, A, that are equivalent to a given present amount, P. This is the converse of the uniform series present amount factor. The equation for the uniform series capital recovery factor is obtained by solving for A in the uniform series present amount factor. That is,

$$A = P\left[\frac{i(1+i)^n}{(1+i)^n - 1}\right]$$

Example 3.5

Suppose a piece of equipment needed to launch a project must be purchased at a cost of $50,000. The entire cost is to be financed at 13.5% per year and repaid on a monthly installment schedule over 4 years. It is desired to calculate what the monthly loan payments will be. It is assumed that the first loan payment will be made exactly 1 month after the equipment is financed. If the interest rate of 13.5% per year is compounded monthly, then the interest rate per month will be 13.5%/12 = 1.125% per month. The number of interest periods over which the loan will be repaid is 4(12) = 48 months. Consequently, the monthly loan payments are calculated to be:

$$A = \$50,000\left[\frac{0.01125(1+0.01123)^{48}}{(1+0.01125)^{48} - 1}\right]$$

$$= \$1353.82$$

3.10　Uniform series compound amount factor

The series compound amount factor is used to calculate a single future amount that is equivalent to a uniform series of equal end-of-period payments. The cash flow is shown in Figure 3.4. Note that the future amount

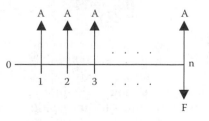

Figure 3.4 Uniform series compound amount cash flow.

occurs at the same point in time as the last amount in the uniform series of payments. The factor is derived as follows:

$$F = \sum_{t=1}^{n} A(1+i)^{n-t}$$

$$= A\left[\frac{(1+i)^n - 1}{i}\right]$$

Example 3.6

If equal end-of-year deposits of $5,000 are made to a project fund paying 8% per year for 10 years, how much can be expected to be available for withdrawal from the account for capital expenditure immediately after the last deposit is made?

$$F = \$5,000\left[\frac{(1+0.08)^{10} - 1}{0.08}\right]$$

$$= \$72,432.50$$

3.11   Uniform series sinking fund factor

The sinking fund factor is used to calculate the uniform series of equal end-of-period amounts, A, that are equivalent to a single future amount, F. This is the reverse of the uniform series compound amount factor. The formula for the sinking fund is obtained by solving for A in the formula for the uniform series compound amount factor. That is,

$$A = F\left[\frac{i}{(1+i)^n - 1}\right]$$

Example 3.7

How large are the end-of-year equal amounts that must be deposited into a project account so that a balance of $75,000 will be available for withdrawal immediately after the 12th annual deposit is made? The initial balance in the account is zero at the beginning of the first year. The account pays 10% interest per year. Using the formula for the sinking fund factor, the required annual deposits are:

$$A = \$75,000 \left[\frac{0.10}{\left(1+0.10\right)^{12} - 1} \right]$$

$$= \$3,507.25$$

3.12   Capitalized cost formula

Capitalized cost refers to the present value of a single amount that is equivalent to a perpetual series of equal end-of-period payments. This is an extension of the series present worth factor with an infinitely large number of periods. This is shown graphically in Figure 3.5.

Using the limit theorem from calculus as n approaches infinity, the series present worth factor reduces to the following formula for the capitalized cost:

$$P = \lim_{n \to \infty} A \left[\frac{\left(1+i\right)^{n} - 1}{i\left(1+i\right)^{n}} \right]$$

$$= A \left\{ \lim_{n \to \infty} \left[\frac{\left(1+i\right)^{n} - 1}{i\left(1+i\right)^{n}} \right] \right\}$$

$$= A \left(\frac{1}{i} \right)$$

There are several real-world investments that can be computed using this idea of capitalized cost formula. These include scholarship funds, maintenance of public buildings, and maintenance of roads and bridges, among others.

Example 3.8

How much should be deposited in a general fund to service a recurring public service project to the tune of $6,500 per year forever if the fund yields

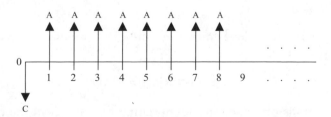

Figure 3.5 Capitalized cost cash flow.

an annual interest rate of 11%? Using the capitalized cost formula, the required one-time deposit to the general fund is as follows:

$$P = \frac{\$6500}{0.11}$$

$$= \$59,090.91$$

Example 3.9

A football stadium is expected to have an annual maintenance expense of $75,000. What amount must be deposited today in an account that pays a fixed interest rate of 12% per year to provide for this annual maintenance expense forever?

The amount of money to be deposited today is

$$\frac{75,000}{0.12} = \$625,000$$

That is, if $625,000 is deposited today into this account, it will pay $75,000 annually forever. This is the power of the compounded interest rate.

3.13   Permanent investments formula

This measure is the reverse of capitalized cost. It is the *net annual value* (NAV) of an alternative that has an infinitely long period. Public projects such as bridges, dams, irrigation systems, and railroads fall into this category. In addition, permanent and charitable organization endowments are evaluated using this approach. The NAV in the case of permanent investments is given by:

$$A = Pi$$

Example 3.10

If we deposit $25,000 in an account that pays a fixed interest rate of 10% today, what amount can be withdrawn each year to sponsor college scholarships forever?

Solution

Using the permanent investments formula, the required annual college scholarship worth is:

$$A = \$25,000 \times 0.10 = \$2,500$$

The formulas presented in the preceding text represent the basic cashflow conversion factors. The factors are tabulated in Appendix E. Variations in the cash-flow profiles include situations where payments are made at the beginning of each period rather than at the end, and situations where a series of payments contains unequal amounts. Conversion formulas can be derived mathematically for those special cases by using the basic factors already presented. Conversion factors for some complicated cash-flow profiles are now discussed.

3.14 Arithmetic gradient series

The gradient series cash flow involves an increase of a fixed amount in the cash flow at the end of each period. Thus, the amount at a given point in time is greater than the amount during the preceding period by a constant amount. This constant amount is denoted by G. Figure 3.6 shows the basic gradient series, in which the base amount at the end of the first period is zero. The size of the cash flow in the gradient series at the end of period t is calculated as follows:

$$A_t = (t-1)G, \quad t = 1, 2, \ldots, n$$

The total present value of the gradient series is calculated by using the present amount factor to convert each individual amount from time t to time 0 at an interest rate of $i\%$ per period and then by summing up the resulting present values. The finite summation reduces to a closed form, as follows:

$$P = \sum_{t=1}^{n} A_t (1+i)^{-t}$$

$$= \sum_{t=1}^{n} (t-1)G(1+i)^{-t}$$

$$= G \sum_{t=1}^{n} (t-1)(1+i)^{-t}$$

$$= G \left[\frac{(1+i)^n - (1+ni)}{i^2 (1+i)^n} \right]$$

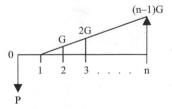

Figure 3.6 Arithmetic gradient cash flow with zero base amount.

Example 3.11

The cost of supplies for a 10-year project increases by $1,500 every year, starting at the end of the second year. There is no supplies cost at the end of the first year. If the interest rate is 8% per year, determine the present amount that must be set aside at time zero to take care of all the future supplies expenditures. We have $G = 1,500$, $i = 0.08$, and $n = 10$. Using the arithmetic gradient formula, we obtain the following:

$$P = 1500 \left[\frac{1 - \left(1 + 10(0.08)\right)\left(1 + 0.08\right)^{-10}}{\left(0.08\right)^2} \right]$$

$$= \$1500 \left(25.9768\right)$$

$$= \$38,965.20$$

In many cases, an arithmetic gradient starts with some base amount at the end of the first period and then increases by a constant amount thereafter. The nonzero base amount is denoted as A_1. Figure 3.7 shows this type of cash flow.

The calculation of the present amount for such cash flows requires breaking the cash flow into a uniform series cash flow of amount A_1 and an arithmetic gradient cash flow with zero base amount. The uniform series present worth formula is used to calculate the present worth of the uniform series portion, and the basic gradient series formula is used to calculate the gradient portion. The overall present worth is then calculated as follows:

$$P = P_{\text{uniform series}} + P_{\text{gradient series}}$$

$$= A_1 \left[\frac{\left(1 + i\right)^n - 1}{i\left(1 + i\right)^n} \right] + G \left[\frac{\left(1 + i\right)^n - \left(1 + ni\right)}{i^2\left(1 + i\right)^n} \right]$$

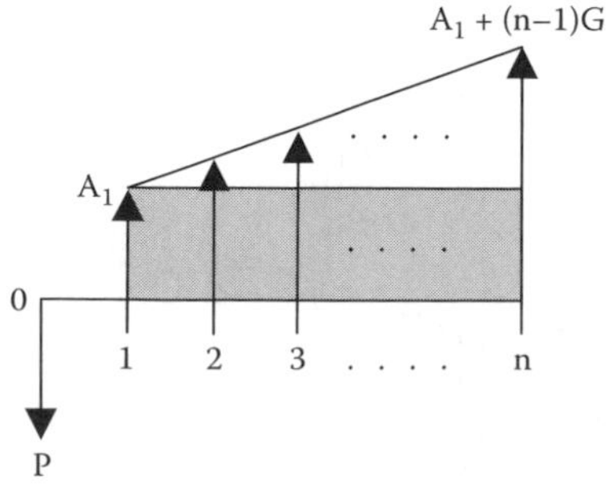

Figure 3.7 Arithmetic gradient cash flow with nonzero base amount.

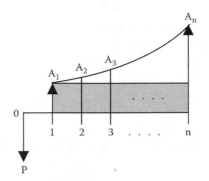

Figure 3.8 Increasing geometric series cash flow.

3.15   *Increasing geometric series cash flow*

In an increasing geometric series cash flow, the amounts in the cash flow increase by a constant percentage from period to period. There is a positive base amount, A_1, at the end of period one. Figure 3.8 shows an increasing geometric series. The amount at time t is denoted as

$$A_t = A_{t-1}(1+j), \quad t = 2,3,\dots,n$$

where j is the percentage increase in the cash flow from period to period. By doing a series of back substitutions, we can represent A_t in terms of A_1 instead of in terms of A_{t-1}, as shown:

$$A_2 = A_1(1+j)$$

$$A_3 = A_2(1+j) = A_1(1+j)(1+j)$$

$$\dots$$

$$A_t = A_1(1+j)^{t-1}, \quad t = 1,2,3,\dots,n$$

The formula for calculating the present worth of the increasing geometric series cash flow is derived by summing the present values of the individual cash-flow amounts. That is,

$$P = \sum_{t=1}^{n} A_t(1+i)^{-t}$$

$$= \sum_{t=1}^{n} \left[A_1(1+j)^{t-1} \right](1+i)^{-t}$$

$$= \frac{A_1}{(1+j)} \sum_{t=1}^{n} \left(\frac{1+j}{1+i} \right)^t$$

$$= A_1 \left[\frac{1-(1+j)^n(1+i)-n}{i-j} \right], \quad i \neq j$$

If $i = j$, the preceding formula reduces to the limit as i approaches j ($i \rightarrow j$), shown as follows:

$$P = \frac{nA_1}{1+i}, \quad i = j$$

Example 3.12

Suppose that funding for a 5-year project is to increase by 6% every year, with an initial funding of $20,000 at the end of the first year. Determine how much must be deposited into a budget account at time zero in order to cover the anticipated funding levels if the budget account pays 10% interest per year. We have $j = 6\%$, $i = 10\%$, $n = 5$, $A_1 = \$20,000$. Therefore,

$$P = 20,000 \left[\frac{1-(1+0.06)^5(1+0.10)^5}{0.10-0.06} \right]$$

$$= \$20,000 \left(4.2267 \right)$$

$$= \$84,533.60$$

3.16 *Decreasing geometric series cash flow*

In a decreasing geometric series cash flow, the amounts in the cash flow decrease by a constant percentage from period to period. The cash flow starts at some positive base amount, A_1, at the end of period one. Figure 3.9 shows a decreasing geometric series. The amount at time t is denoted as follows:

$$A_t = A_{t-1}(1-j), \quad t = 2,3,\ldots,n$$

where j is the percentage decrease in the cash flow from period to period. As in the case of the increasing geometric series, we can represent A_t in terms of A_1:

$$A_2 = A_1(1-j)$$

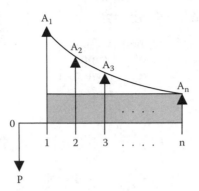

Figure 3.9 Decreasing geometric series cash flow.

$$A_3 = A_2\left(1-j\right) = A_1\left(1-j\right)\left(1-j\right)$$

$$\cdots$$

$$A_t = A_1\left(1-j\right)^{t-1}, \quad t = 1,2,3,\ldots,n$$

The formula for calculating the present worth of the decreasing geometric series cash flow is derived by finite summation, as in the case of the increasing geometric series. The final formula is:

$$P = A_1\left[\frac{1-\left(1-j\right)^n\left(1+i\right)^{-n}}{i+j}\right]$$

Example 3.13

A contract amount for a 3-year project is expected to decrease by 10% every year with an initial contract of $100,000 at the end of the first year. Determine how much must be available in a contract reservoir fund at time zero in order to cover the contract amounts. The fund pays 10% interest per year. Because $j = 10\%$, $i = 10\%$, $n = 3$, $A_1 = \$100,000$, we should have

$$P = 100,000\left[\frac{1+\left(1+0.10\right)^3\left(1+0.10\right)^{-3}}{0.10+0.10}\right]$$

$$= \$100,000\left(2.2615\right)$$

$$= \$226,150$$

3.17   *Internal rate of return*

The IRR for a cash flow is defined as the interest rate that equates the future worth at time n or present worth at time 0 of the cash flow to zero. If we let $i*$ denote the internal rate of return, we then have the following:

$$FW_{t=n} = \sum_{t=0}^{n} \left(\pm A_t \right) \left(1 + i* \right)^{n-t} = 0$$

$$PW_{t=0} = \sum_{t=0}^{n} \left(\pm A_t \right) \left(1 + i* \right)^{-t} = 0$$

where "+" is used in the summation for positive cash-flow amounts or receipts, and "–" is used for negative cash-flow amounts or disbursements. A_t denotes the cash-flow amount at time t, which may be a receipt (+) or a disbursement (–). The value of $i*$ is referred to as the *discounted cash flow rate of return, internal rate of return,* or *true rate of return.* The procedure just discussed essentially calculates the net future worth or the net present worth of the cash flow. That is,

Net Future Worth = Future Worth of Receipts – Future Worth of Disbursements

$$NFW = FW_{(receipts)} - FW_{(disbursements)}$$

Net Present Worth = Present Worth of Receipts – Present Worth of Disbursements

$$NPW = PW_{(receipts)} - PW_{(disbursements)}$$

Setting the NPW or NFW equal to zero and solving for the unknown variable i determines the internal rate of return of the cash flow.

3.18   *Benefit/cost ratio*

The computational methods described previously are mostly used for private projects because the objective of most private projects is to maximize profits. Public projects, on the other hand, are executed to provide services to the citizenry at no profit; therefore, they require a special method of analysis. Benefit/cost (B/C) ratio analysis is normally used for evaluating public projects. It has its origins in the Flood Act of 1936, which requires that, for a federally financed project to be justified, its benefits must, at minimum, equal its costs. B/C ratio is the systematic method of calculating the ratio of project benefits to project costs at a discounted rate. For over 60 years, the B/C ratio method has been the accepted procedure for making "go" or "no-go" decisions on independent and mutually exclusive projects in the public sector.

The B/C ratio of a cash flow is the ratio of the present worth of benefits to the present worth of costs. This is defined as

$$B/C = \dfrac{\displaystyle\sum_{t=0}^{n} B_t \left(1+i\right)^{-t}}{\displaystyle\sum_{t=0}^{n} C_t \left(1+i\right)^{-t}}$$

$$= \dfrac{PW_{\text{benefits}}}{PW_{\text{costs}}} = \dfrac{AW_{\text{benefits}}}{AW_{\text{costs}}}$$

where B_t is the benefit (receipt) at time t and C_t is the cost (disbursement) at time t. If the B/C ratio is greater than one, then the investment is acceptable. If the ratio is less than one, the investment is not acceptable. A ratio of one indicates a break-even situation for the project.

Example 3.14

Consider the following investment opportunity by Knox County.

Initial cost = $600,000
Benefit per year at the end of Years 1 and 2 = $30,000
Benefit per year at the end of Years 3 to 30 = $50,000

If Knox County set the interest at 7% per year, would this be an economically feasible investment for the county?

Solution

For this problem, we will use both the PW and the AW approaches in order to show that both equations would give the same results.

Using Present Worth

$$PW_{\text{benefits}} = 30,000\left(P/A,7\%,2\right) + 50,000\left(P/A,7\%,28\right)\left(P/F,7\%,2\right)$$

$$= \$584,267.16$$

$$PW_{\text{costs}} = \$600,000$$

Using Annual Worth

$$AW_{\text{benefits}} = \left[30,000\left(P/A,7\%,2\right) + 50,000\left(P/A,7\%,28\right)\left(P/F,7\%,2\right)\right]\left(A/P,7\%,30\right)$$

$$= \$47,086.09$$

$$AW_{\text{costs}} = \$600,000\left(A/P,7\%,30\right)$$

$$= \$48,354.00$$

$$B/C = \frac{584,267.16}{600,000.00} = \frac{47,086.09}{48,354.00} = 0.97$$

The B/C ratio indicates that the investment is not economically feasible because B/C < 1.0. However, one can see that the PW and AW methods both produced the same result.

3.19 Simple payback period

The term *payback period* refers to the length of time it will take to recover an initial investment. The approach does not consider the impact of the time value of money. Consequently, it is not an accurate method of evaluating the worth of an investment. However, it is a simple technique that is used widely to perform a "quick-and-dirty" or superficial assessment of investment performance. The technique considers only the initial cost. Other costs that may occur after time zero are not included in the calculation. The payback period is defined as the smallest value of $n(n_{min})$ that satisfies the following expression:

$$\sum_{t=1}^{n_{min}} R_t \geq C$$

where R_t is the revenue at time t and C_0 is the initial investment. The procedure calls for a simple addition of the revenues, period by period, until enough total has been accumulated to offset the initial investment.

Example 3.15

An organization is considering installing a new computer system that will generate significant savings in material and labor requirements for order processing. The system has an initial cost of $50,000. It is expected to save the organization $20,000 a year. The system has an anticipated useful life of 5 years with a salvage value of $5,000. Determine how long it would take for the system to pay for itself from the savings it is expected to generate. Because the annual savings are uniform, we can calculate the payback period by simply dividing the initial cost by the annual savings. That is:

$$n_{min} = \frac{\$50,000}{\$20,000}$$

$$= 2.5 \text{ years}$$

Note that the salvage value of $5,000 is not included in the preceding calculation because the amount is not realized until the end of the useful life of the asset (i.e., after 5 years). In some cases, it may be desirable to consider the salvage value. In that case, the amount to be offset by the annual savings will be the net cost of the asset, represented here as

$$n_{min} = \frac{\$50,000 - \$5000}{\$20,000}$$

$$= 2.25 \text{ years}$$

If there are tax liabilities associated with the annual savings, those liabilities must be deducted from the savings before calculating the payback period. The simple payback period does not take the time value of money into consideration; however, it is a concept readily understood by people unfamiliar with economic analysis.

3.20   Discounted payback period

The discounted payback period is a payback analysis approach in which the revenues are reinvested at a certain interest rate. The payback period is determined when enough money has been accumulated at the given interest rate to offset the initial cost as well as other interim costs. In this case, the calculation is done with the aid of the following expression:

$$\sum_{t=1}^{n_{min}} R_t \left(1+i\right)^{n_{min}-1} \geq \sum_{t=0}^{n_{min}} C_t$$

Example 3.16

A new solar-cell unit is to be installed in an office complex at an initial cost of $150,000. It is expected that the system will generate annual cost savings of $22,500 on the electricity bill. The solar cell unit will need to be overhauled every 5 years at a cost of $5,000 per overhaul. If the annual interest rate is 10%, find the discounted payback period for the solar-cell unit considering the time value of money. The costs of overhaul are to be considered in calculating the discounted payback period.

Solution

Using the single-payment compound amount factor for one period iteratively, the following solution is obtained:

Time	Cumulative Savings
1	$22,500
2	$22,500 + $22,500 (1.10)^1 = $47,250
3	$22,500 + $47,250 (1.10)^1 = $74,475
4	$22,500 + $74,475 (1.10)^1 = $104,422.50
5	$22,500 + $104,422.50 (1.10)^1 – $5000 = $132,364.75
6	$22,500 + $132,364.75 (1.10)^1 = $168,101.23

The initial investment is $150,000. By the end of period 6, we have accumulated $168,101.23, more than the initial cost. Interpolating between period 5 and period 6, we obtain:

$$n_{min} = 5 + \frac{150,000 - 132,364.75}{168,101.25 - 132,364.75}(6 - 5)$$

$$= 5.49$$

That is, it will take 5.49 years, or 5 years and 6 months, to recover the initial investment.

Example 3.17

For the following cash flows:

1. Calculate the simple payback period.
2. Calculate the discount payback period.

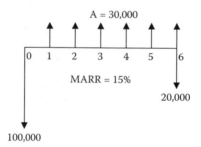

A = 30,000

MARR = 15%

20,000

100,000

Solution

In order to solve this problem, we make a payback table to facilitate easy computation:

EOY	Net Cash Flow	Cumulative PV @ $i = 0$% $= \sum_{k=0}^{n} a_k$	PV @ $i > 0$% (P/F,15%,n)	Cumulative PV @ $i = 15$% $= \sum_{k=0}^{n} b_k$
(n)	(a)		(b)	
0	$100,000	$100,000	$100,000	$100,000
1	$30,000	$70,000	$26,088	$73,912
2	$30,000	$40,000	$22,683	$51,229
3	$30,000	$10,000	$19,725	$31,504
4	$30,000	**$20,000**	$17,154	$14,350
5	$30,000	$50,000	$14,916	**$566**
6	$10,000	$60,000	$4,323	$4,889

Based on the computations in the payback table, the *simple payback period* (SPP) is determined from the third column, and the *discounted payback period* (DPP) is determined from the fifth column. Therefore, the SPP is 4 years, the fourth year being when the cumulative PV becomes a positive value. The DPP is 5 years because the cumulative PV becomes a positive value in the fifth year. It must be noted that, in computing these values, the cash flows after SPP or DPP are not taken into consideration; therefore, both SPP and DPP techniques are usually used for initial screening of potential investment alternatives. They must never be used for final selection without considering other techniques such as Net Present Value and/or Internal Rate of Returns techniques.

3.21   Fixed and variable interest rates

An interest rate may be fixed or may vary from period to period over the useful life of an investment, and companies, when evaluating investment alternatives, need also to consider the variable interest rates involved, especially in long-term investments. Some of the factors responsible for varying interest rates include changes in nominal and international economies, effects of inflation, and changes in market share. Loan rates, such as mortgage loan rates, may be adjusted from year to year based on the inflation index of the U.S. Consumer Price Index (CPI). If the variations in interest rate from period to period are not large, cash-flow calculations usually ignore their effects. However, the results of the computation will vary considerably if the variations in interest rates are large. In such cases, the varying interest rates should be considered in economic analysis, even though such consideration may become computationally involved.

Example 3.18

Find the present worth (present value) of the following cash flows if for $n < 5$, $i = 0.5\%$, and for $n > 4$, $i = 0.25\%$.

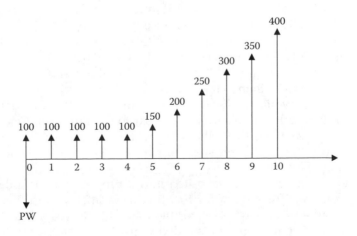

Solution

This is typical of several real-world cash flows. The computation must be carefully done in order to avoid errors.

The present value is given as follows:

$$PV = A_0 + A_{1-4}\left(P/A, 0.5\%, 4\right)$$

$$+\left[A_5\left(P/A, 0.25\%, 6\right) + G\left(P/G, 0.25\%, 6\right)\right]\left(P/F, 0.5\%, 4\right)$$

$$= 100 + 100\left(3.950\right) + \left[150\left(5.948\right) + 50\left(14.826\right)\right]\left(0.9802\right)$$

$$= 100 + 395 + \left[892.2 + 741.3\right]\left(0.9802\right)$$

$$= \$2096.16$$

Therefore, the present value for these cash flows with two different interest rates is $2,096.16. The interest rate for $n > 4$ affects only cash flows in periods 5 to 10, whereas the interest rate for $n < 5$ affects cash flows in periods 1 to 10.

Practice problems for fundamentals of economic analysis

3.1 Calculate how long it would take your current personal or family savings to double at the current interest rate you are being offered by your bank. What will it take for you or your family to reduce the calculated period by one half?

3.2 If the nominal interest rate on a savings account is 0.25% payable, or compounded, quarterly, what is the effective annual interest rate? If $1,000 is deposited into this account quarterly, how much would be available in the account after 10 years?

3.3 A football player signed an $11 million, 10-year contract package with a football team he joined recently. Based on this contract, the football player will receive the following benefits: his yearly salary starts at $300,000 and goes up yearly to $400,000, $500,000, $600,000, $700,000, $1 million, $1.1 million, $1.2 million, $1.3 million and, finally, to $1.4 million in the 10th year. Besides his salary, a $2.5 million bonus is available that will pay him $500,000 immediately and $500,000 each year from the 11th year to the 14th year. Calculate the total present worth of the salaries and bonuses if the prevailing interest rate is 8% compounded per year. Did the footballer get the value of his contract?

3.4 How much must be deposited into a project account today if the project cost for each of the first 5 years is $12,000, and this amount increases by 10% per year for the following 10 years if the account pays 2.5% per year, compounded yearly?

3.5   In order to maintain RAB University's football stadium, the athletic department of the university needs annual maintenance costs of $60,000, annual insurance costs of $5,000, and annual utilities costs of $1,500. In addition, the department needs $100,000 worth of donations every 10 years for expansion projects. If the department opens a special account that pays 2.3% per year, how much must be deposited into this account now to pay for these annual costs forever?

3.6   A newly implemented technology in a manufacturing plant pays zero revenue in the first two years but $1,000 revenue in the third and fourth years; this amount increases by $500 annually for the following 5 years. If the company uses a MARR of 5.5% per year, what is the present value of these cash flows? What is the annual equivalent of the benefits over a 7-year period?

3.7   Repeat Problem 3.6 if the MARR for the company is 5.5% in the first 4 years and increases to 6% starting in year five.

3.8   A company borrowed $10,000 at 12% interest per year compounded yearly. The loan was repaid at $2,000 per year for the first 4 years and $2,200 in the fifth year. How much must be paid in the sixth year to pay off the loan?

3.9   A university alumnus wants to save $25,000 over 15 years so that he could start a scholarship for students in industrial engineering. To have this amount when it is needed, annual payments will be made into a savings account that earns 8% interest per year. What is the amount of each annual payment?

3.10   A small-scale industry thinks that it will produce 10,000 t of metal during the coming year. If the processes are controlled properly, then the metal is going to increase 5% per year thereafter for the next 6 years. Profit per ton of metal is $14 for years 1 to 7. If the industry earns 15% per year on this capital, what is the future equivalent of the industry's cash flows at the EOY 7?

3.11   My grandmother just purchased a new house for $500,000. She made a down payment of 50% of the negotiated price and then makes a payment of $2,000.00 per month for 36 months. Furthermore, she thinks that she can resell the house for $600,000 due to the increasing real estate bubble at the end of 3 years. Draw a cash flow for this from my grandmother's point of view.

3.12   A manufacturing unit in a facility is thinking of purchasing automatic lathes that could save $67,000 per year on labor and scrap. This lathe has an expected life of 5 years and no market value. If the company tells the manufacturing unit that it is expecting to see a 15% ROI per year, how much could be justified now for the purchase of this lathe? Explain it from the manufacturing unit's perspective.

3.13   An ambitious student wants to start a restaurant, so he wants to buy a nice kitchen set consisting of two state-of-the-art grills, three stoves, two dishwashers, three refrigerators, two ovens, and three microwave ovens, along with some other stuff. So he plans on spending

$112,000 on the equipment alone. He feels that these all would produce him a net income of $25,000 per year. If he does not change his mind and keeps this equipment for 4 years considering he doesn't lose much in this business, what would be the resale value of all this equipment at the EOY 4 to justify his investment? A 15% annual return on investment is desired.

3.14   In the process of saving some money for my kid's college, I plan to make six annual deposits of $4000 into a secret savings account that pays an interest of 4% compounded annually. Two years after making the last deposit, the interest rate increases to 7% compounded annually. Twelve years after the last deposit, the accumulated money is taken out for the first time to pay for his tuition. How much is withdrawn?

3.15   A newly employed engineer wants to find out how much he should invest at 12% nominal interest, compounded monthly, to provide an annuity of $25,000 (per year) for 6 years starting 12 years from now.

chapter four

Economic methods for comparing investment alternatives

The objective of performing an economic analysis is to make it easier to decide between or among potential investments that are competing for limited financial resources. Investment opportunities are either mutually exclusive or independent. For mutually exclusive investment opportunities, only one viable project can be selected; therefore, each project competes with the others. On the other hand, more than one project may be selected if the investment opportunities are independent; in such a case, the alternatives do not compete with one another in the evaluation.

There are several methods for comparing investment alternatives: the present value analysis, the annual value analysis, the rate of return analysis, and the benefit/cost ratio analysis. In using these methods, the *do-nothing* (DN) option is a viable alternative that must be considered except when an investment alternative must in any case be selected. The selection of the DN alternative as the accepted project means, of course, that no new investment will be initiated. Three different analysis periods are usually used for evaluating alternatives: equal service lives for all alternatives, different service lives for all alternatives, and infinite service lives for all alternatives. The type of analysis period used may influence the method of evaluation chosen. In Chapter 3, we described each of the fundamental equations used in economic analysis; in this chapter, we present their applications in evaluating single and multiple investments.

4.1 Net present value analysis

The *net present value* (NPV) analysis is the application of some of the engineering economic analysis factors in which the present amount is unknown. It is usually used for projects with equal service lives and can be used for evaluating one alternative, two or more mutually exclusive opportunities, or independent alternatives. The NPV analysis evaluates projects by converting all future cash flows into their present equivalent. The guidelines for using the present-value analysis for evaluating investment alternatives follow:

- *For one alternative:* Calculate NPV at the minimum attractive rate of return (MARR). If NPV ≥ 0, the requested MARR is met or exceeded, and the alternative is economically viable.
- *For two or more alternatives:* Calculate the NPV of each alternative at MARR. Select the alternative with the *numerically largest* NPV value. The numerically largest value indicates a lower NPV of cost cash flows (less negative) or a larger NPV of net cash flows (more positive).
- *For independent projects:* Calculate the NPV of each alternative. Select all projects with a NPV ≥ 0 at the given MARR.

Let

P = cash flow value at the present time period. This usually occurs at time 0.

F = cash flow value at some time in the future.

A = a series of equal, consecutive, and end-of-period cash flow. This is also called annuity.

G = a uniform arithmetic gradient increase in period-by-period cash flow.

n = the total number of time periods, which can be in days, weeks, months, or years.

i = interest rate per time period expressed as a percentage.

Z = face, or par, value of a bond.

C = redemption or disposal price (usually equal to Z).

r = bond rate (nominal interest rate) per interest period.

NCF_t = estimated net cash flow for each year t.

NCF_A = estimated equal amount net cash flow for each year.

n_p = discounted payback period.

The general equation for the present value analysis is

$$NPV = A_0 + A(P/A, i, n) + F(P/F, i, n)$$
$$+ \left[A_1(P/A, i, n) + G(P/G, i, n) \right] + A_1(P/A, g, i, n) \tag{4.1}$$

This equation reduces to a manageable size depending on the cash-flow profiles of the alternatives. For example, for bonds, the NPV analysis equation becomes:

$$NPV = A(P/A, i, n) + F(P/F, i, n) = rZ(P/A, i, n) + C(P/F, i, n) \tag{4.2}$$

Two extensions of NPV analysis are *capitalized cost* and *discounted payback period*.

Example 4.1

A $10,000 bond has a nominal interest rate of 10%, paid monthly, and matures 10 years after it is issued. After the original purchaser has had the bond for 5 years, she needs to sell it. The nominal rate for similar bonds is now 11.4%. What is the present worth measure of the bond to potential buyers?

Solution

From the question,

$$C = Z = \$10,000$$

$$r = \frac{10\%}{12} = 0.83\% \text{ per month}$$

$$i = \frac{11.4\%}{12} = 0.95\% \text{ per month}$$

Therefore, using Equation 4.2:

$$NPV = 83.33\big(P/A, 0.95\%, 60\big) + 10,000\big(P/F, 0.95\%, 60\big)$$

$$= \$9468.68$$

That is, the current selling price of the bond is $9,468.68; therefore, a potential buyer must not pay more than this amount for the bond. The ENGINEA software described in Chapter 12 can be used to obtain values for interest rates not tabulated in Appendix E.

Example 4.2

A new manufacturing technology can be implemented using either of two alternative sources of energy: solar cells or a gas power plant. Solar cells will cost $12,600 to install and will have a useful life of 4 years with no salvage value. Annual costs for maintenance are expected to be $1,400. A gas power plant will cost $11,000 to install, with gas costs expected to be $800 per year. Because the technology will be replaced after 4 years, the salvage value of the gas plant is considered to be zero. At an interest rate of 17% per year, which alternative should be selected using the present value approach?

Solution

In order to solve this problem, Equation 4.1 reduces to $NPV = A_0 + A(P/A, i, n)$. Because one of these sources of energy must be selected, DN is not an alternative.

The *NPV* for solar cells is:

$$NPV_{SC} = -12,600 - 1400(P/A, 17\%, 4)$$

$$= -\$16,440.50$$

The NPV for a gas power plant is:

$$NPV_{GPP} = -11,000 - 800(P/A, 17\%, 4)$$

$$= -\$13,194.60$$

Based on the preceding computation, we must select the **gas power plant** option because it requires a smaller cost in the present time.

Example 4.3

A local waste disposal company is considering two alternatives for a new truck. The most likely cash flows for the two alternatives are as follows:

Model	First Cost	Annual Operating Cost	Annual Income	Salvage Value	Useful Life
A	$60,000	$1,000	$13,000	$10,000	10 years
B	$80,000	$2,000	$15,000	$12,000	10 years

Using the net present value approach and interest rate of 8% per year, which truck should the company buy?

Solution

The cash-flow diagram using the ENGINEA for each of the alternatives is shown below:

Cash-flow diagram for Model A:

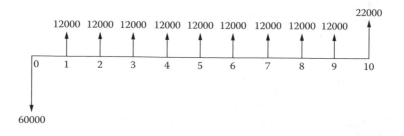

Cash-flow diagram for Model B:

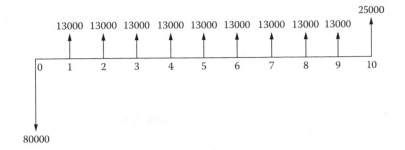

The computation of the NPV follows:

$$NPV_A = -60,000 + 12,000(P/A,8\%,9) + 22,000(P/F,8\%,10) - \$25,152.91$$

$$NPV_B = -80,000 + 13,000(P/A,8\%,9) + 25,000(P/F,8\%,10) = \$12,789.38$$

Therefore, **Model A** should be selected because it has a higher NPV.

Example 4.4

RAB General Hospital is evaluating three contractors to manage its emergency ambulance service for 5 years. Using the present value approach, which of these contractors should be selected if the interest rate is 10% per year?

	Contractor A	Contractor B	Contractor C
Initial cost	$15,000	$20,000	$25,000
Maintenance and operating costs	1,600	1,000	900
Annual benefits	8,000	10,000	13,000
Salvage value	3,000	4,500	6,000

The cash-flow diagrams for Contractors A, B, and C, respectively, are as follows:

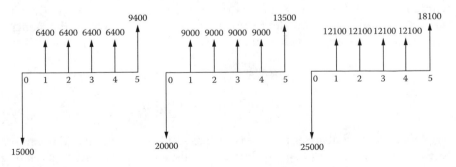

The computation of the NPV follows:

$$NPV_A = -15,000 + 6,400(P/A,10\%,4) + 9,400(P/F,10\%,5) = \$11,123.80$$

$$NPV_B = -20,000 + 9,000(P/A,10\%,4) + 13,500(P/F,10\%,5) = \$16,911.23$$

$$NPV_C = -25,000 + 12,100(P/A,10\%,4) + 18,100(P/F,10\%,5) = \$24,594.05$$

Therefore, Contractor C is the best option for RAB General Hospital.

4.2   Annual value analysis

The *net annual value* (NAV) method of evaluating investment opportunities is the most readily used of all the measures because people easily understand what it means. This method is mostly used for projects with unequal service life because it requires the computation of the equivalent amount of the initial investment and the future amounts for only one project life cycle. NAV analysis converts all future and present cash flows into equal end-of-period amounts. For mutually exclusive alternatives, calculate NAV at MARR and select the viable alternatives based on the following guidelines:

- *One alternative:* Select an alternative with NAV ≥ 0 because MARR is met or exceeded.
- *Two or more alternatives:* Choose alternative with the lowest-cost or the highest-revenue NAV value.

Let
    CR = capital recovery component
    A = annual amount component of other cash flows
    P = initial investment (first cost) of all assets
    S = estimated salvage value of the assets at the end of their useful life
    i = investment interest rate

The annual value amount for an alternative consists of two components: capital recovery for the initial investment P at a stated interest rate (usually at MARR) and the equivalent annual amount A. Therefore, the general equation for the annual value analysis is:

$$NAV = -CR - A$$
$$= -\left[P(A/P,i,n) - S(A/F,i,n)\right] - A \qquad (4.3)$$

NAV is especially useful in areas such as asset replacement and retention, break-even studies and make-or-buy decisions, as well as all studies relating to profit measure. It should be noted that expenditures of money increase

NAV, whereas receipts of money such as selling an asset for its salvage value decrease it. The NAV method assumes the following:

- The service provided will be needed forever because it computes the annual value per cycle.
- The alternatives will be repeated exactly the same in succeeding life cycles. This is especially important when the service life extends several years into the future.
- All cash flows will change by the same amount as the inflation or deflation rate.

The validity of these assumptions is based on the accuracy of the cash-flow estimates. If the cash-flow estimates are very accurate, then the assumptions based on this method will be valid and should minimize the degree of uncertainty surrounding the final decision.

Example 4.5

Repeat Example 4.2 using the annual value approach.

Solution

In order to solve this problem, we use Equation 4.3:

$$NAV = -\left[P(A/P,i,n) - S(A/F,i,n) \right] - A$$

The *NAV* for solar cells:

$$NAV_{SC} = -\left[12,600(A/P,17\%,4) - 0(A/F,17\%,4) \right] - 1400$$

$$= -5992.70$$

The *NAV* for the gas power plant:

$$NAV_{GPP} = -\left[11,000(A/P,17\%,4) - 0(A/F,17\%,4) \right] - 800$$

$$= -4809.50$$

Again, we should select the gas power plant option because it requires less cost on an annual basis.

Example 4.6

An oil-and-gas company now finds it necessary to stop gas flaring. Its E&P (exploration and production) department estimates that the gas processing will cost $30,000 in the first year and will decline by $3,000 each year for the next 10 years. As an alternative, a specialized gas processing company has offered to process the gas for a fixed price of $15,000 per year for the next 10 years, payable at the end of each year. If the oil and gas company uses a 7% interest rate, which alternative must be selected using the annual value analysis?

Solution

The cash-flow diagrams using the ENGINEA software are shown here: Cash-flow diagram based on the E&P department estimates:

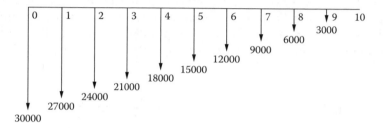

Cash-flow diagram based on the contractor estimates:

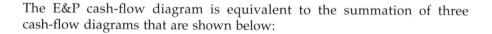

The E&P cash-flow diagram is equivalent to the summation of three cash-flow diagrams that are shown below:

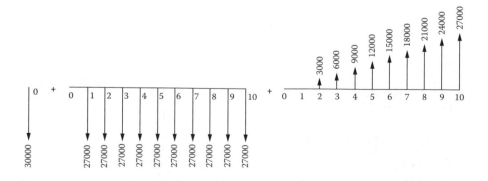

Hence, the NAV computation follows:

$$NAV_{E\&P} = \left(-30{,}000 - 27{,}000\left(P/A,7\%,10\right) + 3{,}000\left(P/G,7\%,10\right)\right)\left(A/P,7\%,10\right)$$

$$= -\$19{,}433.11$$

$$NAV_{Contractor} = -15{,}000\left(A/P,7\%,10\right) - 15{,}000$$

$$= -\$17{,}135.66$$

Therefore, the alternative of using the contractor has the smallest net annual cost and should be selected.

Example 4.7

The project engineer of a food-processing company is considering a new labeling style that can be produced by two alternative machines. The respective costs and benefits of the machines follow:

Cash flow	Machine A	Machine B
Initial cost	$25,000	$15,000
Maintenance costs	400	1,600
Annual benefit	13,000	8,000
Salvage value	6,000	3,000
Useful life	10 years	7 years

If the annual interest rate is 12%, which machine should be selected according to the annual cash-flow analysis?

Solution

The cash-flow diagram for Machine A:

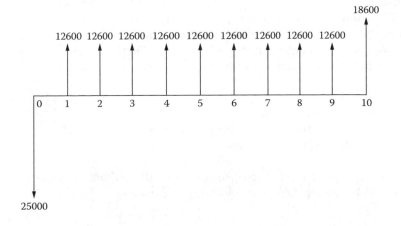

Therefore,

$$NPV_A = -25,000(A/P, 12\%, 10) + 12,600 + 18,600(A/F, 12\%, 10) = \$8517.30$$

The cash-flow diagram for Machine B:

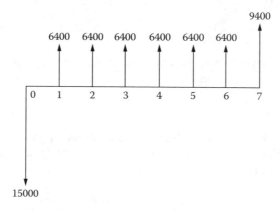

Therefore,

$$NPV_B = -15,000(A/P, 12\%, 7) + 6400 + 9400(A/F, 12\%, 7) = \$3410.59$$

The annual cash-flow analysis shows that Machine A should be selected because its net annual value is greater than that of Machine B.

4.3   Internal rate of return analysis

The internal rate of return (IRR) is the third and most widely used method of measurement in the industry. It is also referred to as simply rate of return (ROR) or return on investment (ROI). It is defined as the interest rate that equates the equivalent value of investment cash inflows (receipts and savings) to the equivalent value of cash outflows (expenditures) — that is, the interest rate at which the benefits are equivalent to the costs.

    Let
        NPV = present value
        $EUAB$ = equivalent uniform annual benefits
        $EUAC$ = equivalent uniform annual costs

If i^* denotes the internal rate of return, then the unknown interest rate can be solved for either using the following expressions:

Table 4.1 Application of Descartes' Rule of Signs to Cash Flows

Number of Sign Changes	Number of Positive Roots
0	No positive roots
1	One positive root
2	Two positive roots or no positive roots
3	Three positive roots or one positive root
4	Four positive roots or two positive roots or no positive roots
5	Five positive roots or three positive roots or one positive root

$$PW\left(\text{Benefits}\right) - PW\left(\text{Costs}\right) = 0$$

$$EUAB - EUAC = 0$$

(4.4)

The procedure for selecting the viable alternatives is:

- If $i^* \geq \text{MARR}$, accept the alternative as an economically viable project.
- If $i^* < \text{MARR}$, the alternative is not economically viable.

When applied correctly, the internal rate of return analysis will always result in the same decision as with NPV or NAV analysis. However, there are some difficulties with IRR analysis: multiple i^*, reinvestment at i^*, and computational difficulty. Multiple i^* usually occurs whenever there is more than one sign change in the cash-flow profile; hence, there is no unique i^* value. This accords with Descartes' rule of signs, which states that "an n-degree polynomial will have at most as many real positive roots as there are number of sign changes in the coefficients of the polynomial." Table 4.1 gives a summary of the number of possible positive roots with respect to the number of sign changes in a cash-flow diagram.

In addition, there may be no real value of i^* that will solve Equation 4.4 even though only real values of i^* are valid in economic analysis. Moreover, IRR analysis usually assumes that the selected project can be reinvested at the calculated i^*, but this assumption is not valid in economic analysis. These difficulties have given rise to an alternative form of IRR analysis called External Rate of Return (ERR) analysis.

Example 4.8

If an investor can receive three payments of $1050 each — at the end of 2, 4, and 6 years, respectively, for an investment of $1500 today, what is the resulting rate of return?

Solution

The cash-flow diagram is shown below:

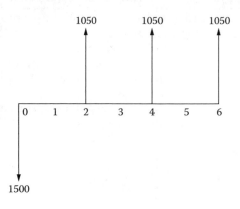

The computation follows:

At what i value is

$$1500 = 1050(P/F,i\%,2) + 1050(P/F,i\%,4) + 1050(P/F,i\%,6)?$$

Try $i = 20\%$:

$$1500 = 1050\left(P/F,20\%,2\right) + 1050\left(P/F,20\%,4\right) + 1050\left(P/F,20\%,6\right)$$

$$1500 = 1050\left(0.6944\right) + 1050\left(0.4823\right) + 1050\left(0.3349\right)$$

$$1500 \neq 1587.15$$

Try $i = 22\%$:

$$1500 = 1050\left(P/F,22\%,2\right) + 1050\left(P/F,22\%,4\right) + 1050\left(P/F,22\%,6\right)$$

$$1500 = 1050\left(0.6719\right) + 1050\left(0.4514\right) + 1050\left(0.3033\right)$$

$$1500 \neq 1497.93$$

The tried i values show that the resulting rate of return lies between 20 and 22%.

$$\frac{20 - X}{20 - 22} = \frac{1587.15 - 1500}{1587.15 - 1497.93}$$

Therefore, $X \approx 21.95\%$. The resulting rate of return is 21.95%.

Example 4.9

A manufacturing plant is for sale for \$240,000. According to a feasibility study, the plant will produce a profit of \$65,000 the first year, but the profit will decline at a constant rate of \$5,000 until it eventually reaches zero. Should an investor looking for a rate of return of at least 20% per year accept this investment?

Solution

The cash-flow diagram of this proposed investment is:

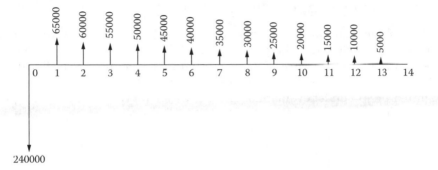

At what i value is $240,000 = 65,000 \left(P/A, i\%, 14 \right) - 5000 \left(P/G, i\%, 14 \right)$?

Try $i = 20\%$:

$$240,000 = 65,000 \left(P/A, 20\%, 14 \right) - 5000 \left(P/G, 20\%, 14 \right)$$

$$= 65,000 \left(4.6106 \right) - 5000 \left(17.6008 \right)$$

$$\neq 211,685$$

Because the NPV for this investment at 20% per year is less than the initial cost of investment, an investor looking for at least a 20% rate of return should not accept this investment. The actual rate of return for this investment is as follows:

Try $i = 15\%$:

$$240,000 = 65,000 \left(P/A, 15\%, 14 \right) - 5000 \left(P/G, 15\%, 14 \right)$$

$$= 65,000 \left(5.7245 \right) - 5000 \left(24.9725 \right)$$

$$\neq 247,220$$

Therefore, the actual rate of return should lie between 15 and 20%:

$$\frac{15-X}{15-20} = \frac{247,200-240,000}{247,200-211,685}$$

$$X \approx 16\%$$

Hence, the actual rate of return is 16%, which is less than the desired 20%. This further supports our initial conclusion that this is not a feasible investment for an investor looking for a rate of return of at least 20%.

4.3.1 External rate of return analysis

The difference between ERR and IRR is that ERR takes into account the interest rate external to the project at which the net cash flow generated or required by the project over its useful life can be reinvested or borrowed. Therefore, this method requires that one know the external MARR for a similar project under evaluation. The expression for calculating ERR is given by

$$F = P\left(1+i'\right)^n \tag{4.5}$$

Using Equation 4.5 requires several steps: (1) the net present value (P) of all net cash outflows is computed at the given external MARR (ε); (2) the net future value (F) of all net cash inflows is computed at the given ε; and (3) these values (P and F) are substituted into Equation 4.5 in order to determine the ERR (i') for the investment. Using this method, a project is acceptable when the calculated i' is greater than ε. However, if i' is equal to ε (a break-even situation), noneconomic factors may be used to justify the final decision. The ERR method has two advantages over the IRR method: it does not resort to simple trial-and-error in determining the unknown rate of return, and it is not subject to the possibility of multiple rates of return even when there are several sign changes in the cash-flow profile.

Example 4.10

A new technology can be purchased today for $25,000, but it will lose $1,000 each year for the first 4 years. An additional $10,000 can be invested in the technology during the fourth year for a profit of $6,000 each year, from the fifth year through the fifteenth year. The salvage value of the technology after 15 years is $31,000. What is the ERR for this investment if ε is 10%?

Solution

For easy computation, we draw the cash-flow diagram for the investment using the ENGINEA software:

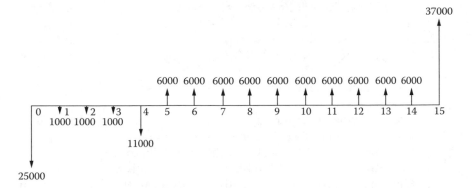

The ERR steps result in the following computation:

1. $P = 25,000 + 1000(P/A,10\%,3) + 11,000(P/F,10\%,4) = \$34,999.90$.
2. $F = 6000(F/A,10\%,11) + 31,000 = \$142.187.20$.
3. $F = P(1 + i')^n = \$142,187.20 = \$34,999.90(1 + i')^{15}$.

Therefore,

$$4.06 - \left(1 + i'\right)^{15}$$

$$\left(4.06\right)^{1/15} = 1 + i'$$

$$1.098 = 1 + i'$$

$$i' = 0.098$$

That is, ERR = 9.8%. This ERR is less than the external MARR; hence, this is not a feasible investment.

4.4   Incremental analysis

Under some circumstances, IRR analysis does not provide the same ranking of alternatives as do NPV and NAV analyses for multiple alternatives. Hence, there is a need for a better approach for analyzing multiple alternatives using the IRR method. Incremental analysis can be defined as the evaluation of the differences between alternatives. The procedure essentially decides whether or not differential costs are justified by differential benefits. Incremental analysis is mandatory for economic analysis involving the use of IRR and benefit/cost (B/C) ratio analyses and when evaluating two or more mutually exclusive alternatives. It is not used for independent projects because more than one project can be selected. The steps involved in using incremental analysis follow.

1. If the IRR (B/C ratio) for each alternative is given, reject all alternatives with an IRR < MARR (B/C < 1.0).
2. Arrange other alternatives in increasing order of initial cost (total costs).
3. Compute incremental cash flow pairwise, starting with the first two alternatives.
4. Compute incremental measures of value based on the appropriate equations.
5. Use the following criteria for selecting the alternatives that will advance to the next stage of comparisons:
   a. If ΔIRR ≥ MARR, select the higher-cost alternative.
   b. If ΔB/C ≥ 1.0, select the higher-cost alternative.
6. Eliminate the defeated alternative and repeat Steps 3 to 5 for the remaining alternatives.
7. Continue until only one alternative remains. This last alternative is the most economically viable one.

4.5 Guidelines for comparison of alternatives

- Total Cash-Flow Approach (Ranking Approach):
  a. Use the individual cash flow for each alternative to calculate the following measures of worth:

    PW, AW, IRR, ERR, Payback Period, or B/C ratio.

  b. Rank the alternatives on the basis of the measures of worth.
  c. Select the highest-ranking alternative as the *preferred* alternative.

- Incremental Cash-Flow Approach:
  Find the incremental cash flow needed to go from a "lower-cost" alternative to a "higher-cost" alternative. Then calculate the measures of worth for the incremental cash flow. The incremental measures of worth are denoted as:

    ΔPW, ΔAW, ΔIRR, ΔERR, ΔPB, ΔB/C

  Step 1. Arrange alternatives in increasing order of initial cost.
  Step 2. Compute incremental cash flow pairwise, starting with the first two alternatives.
  Step 3. Compute incremental measures of worth as explained previously.
  Step 4. Use the following criteria for selecting the alternatives that will advance to the next stage of comparisons:
  - If ΔPW ≥ 0, select higher-cost alternative.
  - If ΔAW ≥ 0, select higher-cost alternative.
  - If ΔFW ≥ 0, select higher-cost alternative.
  - If ΔIRR ≥ MARR, select higher-cost alternative.
  - If ΔERR ≥ MARR, select higher-cost alternative.
  - If ΔB/C ratio ≥ 1.0, select higher-cost alternative.

Step 5. Eliminate the defeated alternative and repeat Steps 2, 3, and 4 for the remaining alternatives. Continue until only one alternative remains. This last alternative is the *preferred* alternative.

Example 4.11

The estimated cash flows for three design alternatives for a laboratory steam-generating plant at a research institute are shown here.

	Alternatives		
	A	B	C
Capital investment	$85,600	$93,200	$71,800
Expected annual savings	$10,200	$15,100	$12,650
Useful life	10	8	9

If MARR is 4% per year, compounded yearly, which alternative, if any, should be selected, using the ROR method if the proposed project is expected to last for only 8 years?

Solution

For this problem, DN is a valid alternative because the question says "if any." If the question had said "must" instead of "if any," then, DN would not be a valid alternative. In order to use the ROR method, it must be the incremental ROR method. Therefore, we use the "incremental ROR" method.

Step 1: Rank the alternatives in order of increasing capital investment (initial cost). For this problem, the ranking is as follows: (1) DN, (2) C, (3) A, and (4) B.

Steps 2 to 5: We now compute the incremental ROR and make a selection:

	C – DN	A – C	B – C
Changes in investment capital	–$71,800	–$13,800	–$21,400
Changes in annual savings	$12,650	–$2,450	$2,450
Incremental IRR	8.34%	<0%	0.70%
Is incremental cost justified?	Yes	No	No
Therefore, select	C	C	C

The best option is alternative C.

Note: The incremental IRR is computed by substituting the PV of Benefits (Changes in Annual Savings) and PV of Cost (Changes in Investment Capital) into Equation 4.4 and finding the value of i that will satisfy the equation.

Example 4.12

There are three alternatives for a public-works project, as given here. If your county *must* do one of these (i.e., DN is not an alternative), which one should it be?

| | Projects | | |
Annual Benefits	A	B	C
Flood reduction	$200,000	$550,000	$650,000
Irrigation	—	175,000	175,000
Recreation	—	45,000	45,000
Annual Costs			
Capital recovery	$293,750	$850,000	$1,143,750
O & M	80,000	50,000	130,000
Useful life	40	43	41

Solution

We must use incremental analysis because the alternatives are public-works projects. The ranking for the projects is A, B, and C.

$$\Delta B/\Delta C(B-A) = \frac{AV(\text{Benefits})}{AV(\text{Costs})} = \frac{770,000 - 200,000}{900,000 - 373,750} = 1.08$$

Select Project B because its incremental B/C ratio is greater than 1.0.

$$\Delta B/\Delta C(C-B) = \frac{AV(\text{Benefits})}{AV(\text{Costs})} = \frac{870,000 - 770,000}{1,273,750 - 900,000} = 0.27$$

Select project B since its incremental B/C ratio is less than 1.0.

Example 4.13

A firm is considering relocating its manufacturing plant from the U.S. to another country. The firm's industrial engineering department, which was given the job of identifying the various alternatives, examined five likely countries together with the costs and benefits of relocating the plant to each. Their findings follow:

Plant Location	First Cost ($1000s)	Uniform Annual Benefit ($1000s)
Mexico	200	30
China	300	95
India	450	160
Pakistan	480	185
Indonesia	500	190

The firm is using a 10-year analysis period, and the annual benefits are expected to be constant over this period. If the firm MARR is 12% per year, where must the manufacturing plant be located, according to analysis by the ROR technique?

Solution

Because the plant must be relocated, the current location (do-nothing alternative) is not an option. Using the incremental method, the solution to this problem using the ENGINEA software is as follows.

The first part of the ENGINEA software computes the rate of return for each alternative:

	Mexico	China	India	Pakistan	Indonesia
Rate of return	8.14	29.23	33.59	36.87	36.28
Retain?	No	Yes	Yes	Yes	Yes

The preceding table shows that only Mexico has a rate of return that is less than MARR and should not be considered in the incremental analysis. Without using the incremental analysis, we can easily (but wrongly) say that Pakistan has the highest ROR and should be selected; however, application of incremental analysis proves otherwise.

	India to China	Pakistan to India	Indonesia to Pakistan
Rate of return	42.04	83.14	21.41
Incremental justified?	Yes	Yes	Yes
Select	India	Pakistan	Indonesia

The results of the incremental analysis show that Indonesia is the most economically feasible location for the manufacturing plant. A portion of the solution using the ENGINEA software is shown here.

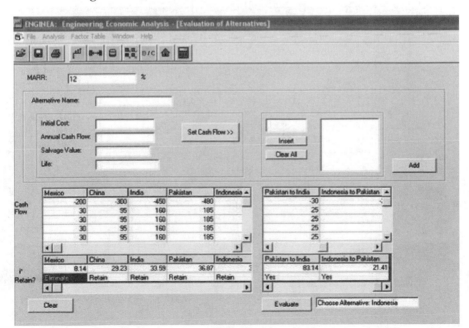

Example 4.14

An oil-and-gas company is considering purchasing a portable offshore drilling rig from three contractors.

	Contractors		
	EC	RE	DE
First cost	$75,000	$125,000	$220,000
Annual benefit	$28,000	$ 43,000	$ 79,000
Maintenance costs	$ 8,000	$ 13,000	$ 38,000
Salvage value	$ 3,000	$ 6,900	$ 16,000

The company uses a 10-year analysis period and a MARR of 15%. Using the incremental ROR analysis, from which contractor should the company purchase the drilling rig?

Solution

Based on the solution using the ENGINEA software, the company should purchase the portable drilling rig from Contractor RE. Contractor DE has an ROR lower than the company MARR. The incremental analysis is performed for Contractor EC and RE's cash flows; the $50,000 increment in initial cost is justified with an ROR of 15.6%. Again, looking at the individual ROR, we can easily (and wrongly) select Contractor EC because it has the highest individual rate of return (23.6%), whereas Contractor RE has a rate of return of 20.4%.

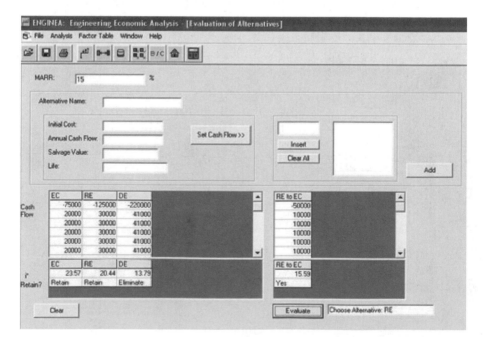

This chapter has presented basic and common methods of comparing investment alternatives. In the next chapter, the comparison techniques are extended to applications in asset retention and replacement analysis.

Practice problems: Comparing investment alternatives

4.1 A $10,000 bond has an interest rate of 8% paid quarterly and matures 10 years after it is issued. The original buyer of this bond decided to sell it after she had it for 6 years. If the nominal rate for similar bonds is now 12% paid quarterly, what is the current selling price?

4.2 Which of the following alternatives should a service company invest in using incremental ROR if MARR is 10% per year? All salvage values are negative, representing cash outflows.

	A	B	C
Initial investment	$80,000	$130,000	$180,000
Annual savings	$25,000	$25,000	$35,000
Salvage	$15,000	$20,000	$20,000
Useful life (years)	10	10	10

4.3 Assume that there are four mutually exclusive options for implementing a public project. Which one must be chosen if the assumed life of the project is 50 years and the interest rate is 7% per year? All data are in $1,000.

	Costs		Benefits		
Option	Initial Construction	Annual Maintenance	Annual Flood	Annual Fire	Annual Recreation
A	$3,500	$55	$500	$50	$75
B	4,200	80	650	$110	156
C	7,000	120	720	$160	75
D	11,000	170	810	$195	179

4.4 There are three alternatives for the construction of a public road in a county. Which of these should be selected, if any?

	Project		
Annual Benefits	A	B	C
Flood reduction	$250,000	$600,000	$700,000
Irrigation	10,000	0	200,000
Recreation	0	175,000	40,000
Annual Costs			
Capital recovery	$387,000	$763,000	$1,250,000
Maintenance	40,000	64,000	98,650

4.5   As the industrial engineer for your team, you are asked to advise management if your company should invest in a project with an expected life of 10 years if the initial investment is $100,000 and your company MARR is 10.5%. The project has a 50% chance it will have annual expenses of $24,000 and a 50% chance it will have annual expenses of $43,000. There is a 30% chance that annual revenue will be $55,000 and a 70% chance that it will be $75,000. What will be your recommendation based on expected present worth of the project?

4.6   An investment has an initial investment of $500, annual revenue of $175 for 5 years, and a salvage value of $100 at the end of year 5. If MARR is 15%, what is ERR, the payback period, and the discounted payback period?

4.7   The management of Indian Chemicals is considering adding an SO_2 scrubber unit in their fertilizer company to remove (SO_2) from stack gases so that cit could be recycled and reused to produce necessary sulfuric acid. You are the industrial engineer on their team, and the mechanical design team presented you with four possible designs. The management of the company is not bothered about the pollution effect but would like to invest in any good design if they are paid a 10% annual rate of return on investment before taxes. Assuming that they would earn a 10% annual ROI, give them your advice on which design would be most profitable to the company. Tabulated here are your calculations for each of the designs. Which, if any, would you recommend to your management and why? The life of each design is 10 years.

4.8   Quality Excellence Airline (QEA) is a small freight line started with very limited capital to serve independent petroleum operators in the arid parts of the Sahara Desert. All of its planes are identical. QEA has been contracting its overhaul work to Chad for $37,000 per plane per year. QEA estimates that, by building a $475,000 maintenance facility with a life of 15 years and a residual (market) value of $115,000, at the end of its life they could handle their own overhaul at a cost of only $27,000 per plane per year. What is the minimum number of planes they must operate to make it economically feasible to build the facility? MARR is 10% per year.

4.9   The following are alternatives that describe possible projects for the use of a vacant plot. In each case, the project cost includes the purchase price of the land.

	Parking Lot	Gas Station
Investment cost	$60,000	$120,000
Annual income	$40,000/year	$95,000/year
Operating expense	$30,000/year	$80,000 in year 1 then increasing by $2000/year
Salvage value	$20,000	$65,000
Useful life	5 years	10 years

If the minimum attractive rate of return is 10%, what should be done, if any, with the land?

4.10 Given the following information about possible investments, what is the best choice at a MARR of 15%? Use the annual worth to make decisions.

	A	B
Investment cost	$10,000	$12,000
Annual benefits	$1500	$900
Useful life	10 years	12 years

chapter five

Asset replacement and retention analysis

Asset replacement, also referred to as retention analysis, is a commonly performed economic analysis in industry. It is an application of the annual value analysis. A replacement occurs when an asset is withdrawn from service for whatever reasons, and another asset is acquired in its place to continue providing the required services. This is also called a like-for-like exchange; therefore, it is assumed that no gain or loss is realized and there is no tax paid or credit received on the exchange. A retirement or disposal, on the other hand, occurs when an asset is salvaged or otherwise disposed of, and the service rendered by the asset is discontinued. In such a case, a gain or loss may be realized, resulting in a tax credit or liability. Some of the reasons for replacing an asset are:

1. Replacement due to inadequacy
2. Replacement due to deterioration
3. Replacement due to obsolescence
4. Replacement due to decline in market value
5. Replacement due to a reduced value to the owner
6. Replacement due to a lower level of desirability of the asset
7. Replacement due to a reduction in product capacity of the asset

5.1 Considerations for replacement analysis

Several factors need to be taken into account in evaluating the replacement of an asset on the basis of the preceding and other reasons. The important factors are described here:

- Deterioration factor: Changes that occur in the physical condition of an asset as a result of aging, unexpected accident, or any other factors that affect the physical condition of the asset.

- Requirements factor: Changes in production plans that affect the economics of use of the asset. For example, the production capacity of an asset may become smaller as a result of plant expansion.
- Technological factor: The impact of changes in technology can also be a factor. For example, the introduction of flash discs affects assets that are used for producing zip discs.
- Financial factor: For example, the lease of an asset may become more financially attractive than ownership; similarly, outsourcing the production of a product may become more attractive than in-house production.

Whatever the reason for a replacement evaluation, it is usually designed to answer the following fundamental question: "Should we replace the current asset (defender) now, or should we keep it for one (or more) additional years."

Therefore, the question is not *whether* the asset should be replaced (because it would be replaced eventually), but *when* it should be replaced. Hence, there are basically two alternatives in a replacement analysis: replacing the defender now or keeping the defender for another year. This type of evaluation study is different from the other studies considered in Chapter 4, in which all the alternatives considered were new. The steps involved are designed to evaluate all the mutually exclusive possible replacement assets, using the same techniques described in Chapter 4, to select the most economically feasible possible replacement (the "challenger"), and to compare the challenger with the "defender," the currently owned asset.

5.2   Terms of replacement analysis

Replacement analysis involves several terms, defined below:

- **Defender:** The currently installed asset being considered for replacement.
- **Challenger:** The potential replacement.
- **Defender first cost:** The current market value of the defender, which is the correct estimate for this term in the replacement study. However, if a defender must be upgraded to make it equivalent to the challenger, the cost of such an upgrade is added to the market value to obtain the correct estimate for this term.
- **Challenger first cost:** The amount that must be recovered when replacing a defender with a challenger. This may be equal to the first cost of the challenger. However, if trade-in is involved, it will be the first cost minus the difference between the trade-in value and the market value of the defender. For example, let us assume we bought an asset (the defender) 5 years ago for $60,000 and that a fair-market value for it today is $30,000. A new asset can be purchased for $50,000

today, and the seller offers a trade-in of $40,000 on the current asset; the true investment in the challenger will be $50,000 – ($40,000 – $30,000) = $40,000. Therefore, the challenger first cost is $40,000. Note that the initial cost of the defender ($60,000) is not relevant to this computation.

- **First cost:** The total cost of preparing the asset for economic use. This includes the purchase price, the delivery cost, the installation cost, and any other costs that must be incurred before the asset can be put into production. This is also called the Basis.
- **Sunk cost:** This is the difference between an asset's Book Value (BV) and its Market Value (MV) at a particular period. Sunk costs have no relevance to the replacement decisions and must be disregarded.
- **Outsider viewpoint:** The perspective that would have been taken by an impartial third party in order to establish a fair MV for the installed asset. The outsider viewpoint forces the analyst to focus on the present and future cash flows in a replacement study, hence avoiding the temptation to dwell on past (sunk) costs.
- **Asset life:** The life of an asset can be seen as divided into three aspects: ownership life, useful life, and economic life. The ownership life of an asset is the period between an owner's acquisition of the asset and that owner's disposal of it. The useful life, on the other hand, is the period during which an asset is kept in productive service. Finally, the economic service life of an asset is the number of periods that results in the minimum Equivalent Uniform Annual Cost (EUAC) of owning and operating the asset. Economic life is often shorter than useful life, and it is usually only 1 year for the defender. Of all these, only economic life is relevant to replacement computations.
- **Marginal cost:** This is the additional cost of increasing production output by one additional unit using the current asset. This is a useful cost parameter in replacement analysis.
- **Before-tax and after-tax analysis:** Replacement analysis can be based on before-tax or after-tax cash flows; however, it is always better to use after-tax cash flows in order to account for the effect of taxes on replacement decisions.

5.3 *Economic service life (ESL)*

Of all the various forms of asset life, ESL is the most important for an asset replacement analysis. This parameter is also called the *minimum cost life*. This is the number of remaining periods that result in the minimum equivalent annual cost of owning and operating an asset. This value is not usually known; therefore, it must be determined in order to perform a replacement analysis. The ESL is determined by calculating the total Annual Value (AV) of costs for the years the asset is in useful service.

5.4   Replacement analysis computation

There are two views to doing computations for replacement analysis as shown below:

- Insider approach (cash-flow approach)
  ⇒ The market value of an existing asset is treated as a reduction in the initial cost of the replacement asset (trade-in-value).
- Outsider approach
  ⇒ The market value of the existing asset is treated as its initial cost (i.e., opportunity cost of keeping the existing asset).

Both views yield the same conclusion regarding replacing or not replacing an asset.

The new asset is considered as the Challenger.
The existing asset is considered as the Defender.

The following example illustrates the computational process.

Let i = 15%

Data for Old Machine
Present worth = \$3600
Worth 1 year from now = \$2800 (salvage value)
Operating cost = \$6000 per year

Cash flow for outsiders View:

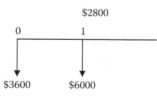

Therefore, the cost of keeping the old machine one more year = EAC.

$$\text{EAC} = (PS)(A \mid P,15\%,1) + S(i) + 6000$$

$$= (36002800)(1.15) + 2800(0.15) + 6000 = \$7340$$

Data for New Machine
Initial cost = \$14,000
Operating Cost = \$4500/year
Salvage value 4 years from now = \$3000

Cash flow for outsider's view:

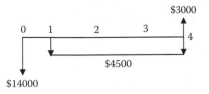

$$EAC = (140003000)(A|P,15\%,4) + 3000(0.15) + 4500 = \$8803$$

Therefore, the new machine will cost more next year. Do not replace.

Insider's View
Alternative 1: Keep the old machine:

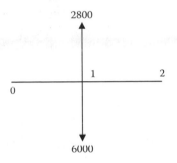

Therefore, the net cost of keeping the old machine one more year = $6000 − $2800 = $3200.

Alternative 2: Buy a new machine:

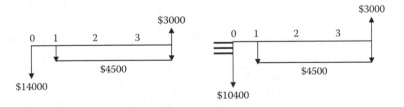

$$EAC = (10,4003000)(A | P,15\%,4) + 3000(0.15) + 4500 = \$7542.22$$

Therefore, keep the old machine one more year.

The optimal replacement time, t^*, can be determined graphically as shown by plotting A vs. CR and evaluating against EAC. The minimum point on the EAC curve corresponds to the optimal replacement time.

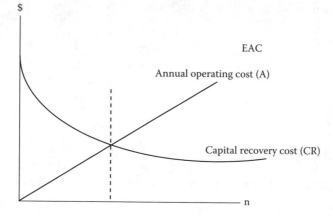

The replacement analysis computations can be divided into three parts: when marginal cost is available and increasing, when it is available but not increasing, and when it is not available at all. The process involved can be depicted as shown in Figure 5.1.

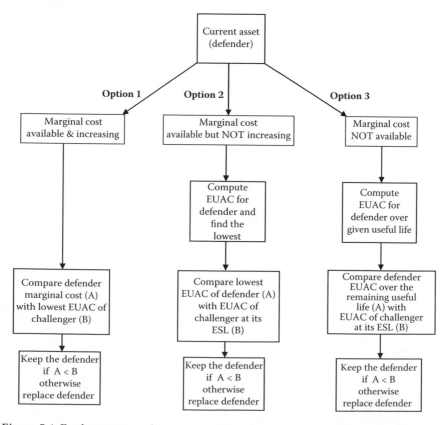

Figure 5.1 Replacement analysis computational process map.

The replacement analysis computational process map shows that the computations required are fairly easy if the marginal cost of the defender is available and such cost is increasing, but they are slightly more complex if the cost is decreasing or the marginal cost is not available.

Example 5.1

A new machine is expected to have the following market values and annual expenses. If the cost of capital is 10%, what is the ESL and minimum EUAC of this machine?

EOY	Market Value	Annual Expenses
0	$100,000	—
1	$80,000	$5,000
2	$60,000	$10,000
3	$45,000	$20,000
4	$30,000	$30,000
5	$20,000	$40,000

Solution

We will use the ENGINEA to solve this problem; however, we will also explain the steps behind the computations. A snapshot of the computation is shown here:

Replacement Analysis 1

Challenger

Cost of Capital   10 %   Select
End of Year (EOY):   6

Enter MV / AE
Market Value (MV) at EOY:
Annual Expenses:
Add

Enter TMC
Total Marginal Costs (TMC)

End of Year (EOY)	Market Value (MV)	Loss in MV	Cost of Capital	Annual Expenses	TMC	EUAC
0	100000	0	0	0	0	0
1	80000	20000	10000	5000	35000	35000
2	60000	20000	8000	10000	38000	36429
3	45000	15000	6000	20000	41000	37810
4	30000	15000	4500	30000	49500	40329
5	20000	10000	3000	40000	53000	42404

Remove Last   Clear All

Economic Service Life (ESL):   1   years

Minimum EUAC: 35000

Result

Steps involved in computation:

1. State whether the asset is a challenger or a defender using the pull-down menu. Because the question says "a new machine," this is the challenger.
2. Select the interest rate as given in the question. The interest rate in this question is 10%.
3. Select whether the given data is based on MV and annual expenses or total marginal cost. In this case, we are given the MV and the annual expenses.
4. Add the MV and annual expenses at the end of each year.

The equations behind each column are shown in the following ENGINEA snapshot:

EOY, k	MV	Loss in MV	Cost of Capital	Annual Expenses	Total Marginal Cost (TMC)	EUAC
(a)	(b)	(c)	(d)	(e)	(f)	(g)

Column (a) = Given
Column (b) = Given
Column (c) = $MV_{(k-1)} - MV_{(k)}$
Column (d) = $(MV_{(k-1)}) \times (i\%)$
Column (e) = Given
Column (f) = [(c) + (d) + (e)], if not given
Column (g) = $EUAC_k$ =

$$\left[\sum_{j=1}^{k} TMC_j \left(P/F, i\%, j \right) \right] \left(A/P, i\%, k \right)$$

Based on the computation, the ESL for the machine is 1 year with a minimum EUAC of \$35,000.

Example 5.2

An existing machine has a Total Marginal Cost (TMC) per year as follows for an interest rate equaling 10% per year. When would you recommend replacing this machine with the machine described in Example 5.1?

EOY	TMC
1	\$60,000
2	\$50,000
3	\$30,000
4	\$55,000
5	\$65,000

Solution

Because the TMC is given, we have to determine whether it is increasing or not. Based on the data given, the TMC is not increasing; therefore, we should compare the minimum EUAC of the defender with the EUAC of the challenger in order to determine when the defender should be replaced (option 2). To do this, we compute the EUAC for the defender.

We will use the ENGINEA to solve this problem, as well. The steps involved are similar to those described in Example 5.1:

1. This is an existing machine; therefore, we select "Defender" as the name of the machine.
2. The interest rate is still 10%.
3. Because the defender information is based on total marginal cost, we select TMC as the basis for the computation.
4. Finally, we add the TMC at the end of each year, remembering that the TMC for year 0 is zero.

A snapshot of the computation table from the ENGINEA software is shown here:

End of Year (EOY)	Market Value (MV)	Loss in MV	Cost of Capital	Annual Expenses	TMC	EUAC
0	0	0	0	0	0	0
1	0	0	0	0	60000	60000
2	0	0	0	0	50000	55238
3	0	0	0	0	30000	47613
4	0	0	0	0	55000	49205
5	0	0	0	0	65000	51792

Economic Service Life (ESL): 3 years

Minimum EUAC: 47613

The computations show that the ESL for the defender is 3 years, with a minimum EUAC of $47,613. The criteria for determining when to replace the defender are that we should keep the defender if the minimum EUAC of the defender is less than the minimum EUAC of the challenger; otherwise, it should be replaced now. Because $47,613 > $35,000, we replace the defender now. Using the ENGINEA software, we click on the "Result" button for the decision. A snapshot of the decision note using the ENGINEA software is shown here:

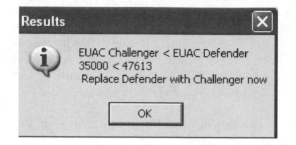

Example 5.3

An existing machine has the TMC per year shown as follows. What is the proper replacement decision for this machine when it is compared to the machine described in Example 5.1 at a 10% interest rate?

EOY	TMC
1	$25,000
2	$29,000
3	$32,000
4	$36,000
5	$39,000

Solution

To answer this question we have to determine whether the given TMC is increasing or not. Based on the data given, the TMC is increasing; therefore, we should compare the TMC of the defender with the minimum EUAC of the challenger in order to determine when the defender should be replaced (option 1). We do not need to compute the EUAC for the defender. Recall that the minimum EUAC for the challenger is $35,000. Comparing this value to the TMC, we see that the defender should be kept for 3 years, after which it should be replaced with the challenger because the TMC for the defender in each of the first 3 years is less than the minimum EUAC for the challenger. However, an extra cost of only $1,000 ($36,000 – $35,000) would be incurred if the defender were to be kept for 4 more years instead of 3 more years. This decision is based on the assumption that the data given will remain valid for the next 3 years and that the challenger would still be available to replace the defender. However, a more appropriate approach is to keep the defender for another year and revisit the study in the following year. This approach would allow for new and possibly additional information for the analysis.

The problems described in the preceding text are based on the assumption that the TMC of the defender is available or can be computed. However, in some cases, this may not be possible. In these cases, the replacement analysis can be done from either the perspective of opportunity cost or the perspective of cash flow. Both perspectives will give the same results when the useful life of the defender and the challenger is the same; however, they will give different and inconsistent results when the useful lives of the defender and challenger are different. The opportunity cost perspective, however, gives consistent results in both cases and is usually used. We will use this perspective to analyze a replacement problem when the marginal cost is not available in the following example.

Example 5.4

For a replacement analysis, the MV of an asset (defender) bought 6 years ago for $900 is $300, and the first cost of a potential challenger is $1200. The remaining life for the defender is 4 years, and the useful life for the challenger is also 4 years. What is the replacement decision if MARR is 10% per year?

Solution

The first cost for the defender is the current MV, which is $300, and the first cost for the challenger is $1200. We determine the AV of the assets over their remaining and useful lives, respectively:

$$AV_{defender} - \$300(A/P,10\%,4) = \$94.64$$

$$AV_{challenger} = \$1,200(A/P,10\%,4) = \$378.56$$

The difference in AV is $283.92, which indicates that the additional cost of the challenger over the next 4 years is justified; therefore, we replace the defender now.

Example 5.5

Repeat Example 5.4, except that, here, the remaining life of the defender is 4 years but the useful life of the challenger is 8 years.

Solution

$$AV_{defender} = \$300(A/P,10\%,4) = \$94.64$$

$$AV_{challenger} = \$1,200(A/P,10\%,8) = \$224.93$$

The difference in AV now is $130.29, which still indicates that the additional cost of the challenger over the next 8 years is justified; therefore, we replace the defender now.

The problems described in the preceding text deal with replacement analysis; however, in some cases it may be desirable to abandon the asset without a replacement. In such cases, we need to determine the most economic period in which to abandon the asset. This is similar to determining the economic life of an asset in that we are determining the span of economic life that maximizes profits or minimizes costs. The following example illustrates the procedures in abandonment problems.

Example 5.6

A small commercial pump is being considered for abandonment. The pump has the following net cash flows and estimated MV (abandonment value of pump) over its useful life. Determine the optimum period for the pump to be abandoned if it was acquired for $5500 and its useful life should not be more than 4 years. MARR for the firm is 10% per year.

	End of Year			
	1	2	3	4
Annual benefits	$500	$600	$750	$900
Estimated MV	$4000	$3500	$3000	$2500

Solution

In order to solve this problem, we compute the present value of keeping the pump for 1, 2, 3, or 4 years. The year that has the maximum present value is the optimum year of abandonment.

If we keep the pump for 1 year:

$$PV_1 = -\$5500 + (\$500 + \$4000)(P/F, 10\%, 1)$$

$$= -\$1409.05$$

If we keep it for 2 years:

$$PV_2 = -\$5500 + \$500(P/F, 10\%, 1) + (\$600 + \$3500)(P/F, 10\%, 2)$$

$$= -\$1657.21$$

If we keep it for 3 years:

$$PV_3 = -\$5500 + \$500(P/F, 10\%, 1) + \$600(P/F, 10\%, 2)$$

$$+ (\$750 + \$3,000)(P/F, 10\%, 3)$$

$$= -\$1732.24$$

If we keep it for 4 years:

$$PV_1 = -\$5500 + \$500(P/F, 10\%, 1) + \$600(P/F, 10\%, 2)$$

$$+\$750(P/F, 10\%, 3) + (\$900 + \$2500)(P/F, 10\%, 4)$$

$$= -\$1663.94$$

Based on the computations, keeping the pump for 1 year minimizes costs; therefore, the optimum period to abandon the pump would be after 1 year.

Practice problems for replacement and retention analysis

5.1 A new machine is expected to have the following MV and annual expenses. If the cost of capital is 12% per year, what is the expected life and minimum EUAC of this machine?

EOY	MV	Annual Expenses
0	$150,000	$5,000
1	$85,000	$10,000
2	$73,000	$18,000
3	$55,000	$29,000
4	$30,000	$40,000
5	$20,000	$55,000

5.2 An existing machine has the following TMC per year. When do you think it is most appropriate to replace this machine with the new machine described in Problem 5.1?

EOY	TMC
1	$55,000
2	$40,000
3	$55,000
4	$48,000
5	$60,000

5.3 Repeat Problem 5.2 if the TMC per year is as follows:

EOY	TMC
1	$45,000
2	$50,000
3	$65,000
4	$78,000
5	$80,000

5.4 An old machine was installed 3 years ago at a cost of $100,000, and you still owe $40,000 before it is completely paid off. It has a present realizable market value of $40,000. If kept, it can be expected to last 5 more years with annual expenses of $10,000 and an MV of $20,000 at the end of 5 years. This machine can be replaced with an improved version costing $125,000 with an expected life of 10 years. The challenger will have estimated annual expenses of $4,200 and a salvage value of $25,000 at the end of 10 years. The machine is expected to be needed indefinitely, and the effect of taxes should be ignored. Using MARR of 20% per year, determine whether to keep or replace the old machine.

5.5 Determine the economic life of a new piece of equipment if it initially cost $10,000. The first-year maintenance cost is $1,500. Maintenance cost is projected to increase $500 per year for each year after the first. Complete the table below if the interest rate is 15% per year.

EOY	Maintenance Cost	EUAC of Capital Recovery	EUAC of Maintenance	Total EUAC
1	$1,500			
2	2,000			
3	2,500			
4	3,000			
5	3,500			
6	4,000			
7	4,500			
8	5,000			
9	5,500			
10	6,000			

chapter six

Depreciation methods

The concept of depreciation is important in economic analysis primarily because it is useful for income-tax computation purposes; however, it has an important role to play in economic analysis for other purposes as well. Depreciation is used in relation to tangible assets such as equipment, computers, machinery, buildings, and vehicles. Almost everything (except land, which is considered a nondepreciable asset) depreciates as time proceeds. Depreciation can be defined as:

- A decline in the market value (MV) of an asset (deterioration).
- A decline in the value of an asset to its owner (obsolescence).
- The allocation of the cost of an asset over its depreciable or useful life. Accountants usually use this definition, and it is adopted in economic analysis for income-tax computation purposes. Therefore, depreciation is a way to claim, over time, an already paid expense for a depreciable asset.

For an asset to be depreciated, it must satisfy these three requirements:

1. The asset must be used for business purposes to generate income.
2. The asset must have a useful life that can be determined, and that is longer than 1 year.
3. The asset must be one that decays, gets used up, wears out, becomes obsolete, or loses value to the owner over time as a result of natural causes.

6.1 Terminology of depreciation

- **Depreciation:** The annual depreciation amount, D_t, is the decreasing value of the asset to the owner. It does not represent an actual cash flow or actual usage pattern.

- **Book depreciation:** This is an internal description of depreciation. It is the reduction in the asset investment due to its usage pattern and expected useful life.
- **Tax depreciation:** This is used for after-tax economic analysis. In the U.S. and many other countries, the annual tax depreciation is tax deductible using the approved method of computation.
- **First cost or unadjusted basis:** This is the cost of preparing the asset for economic use. This is also called the *basis*. This term is used when an asset is new. Adjusted basis is used after some depreciation has been charged.
- **Book value:** This represents the difference between the basis and the accumulated depreciation charges at a particular period. It is also called the undepreciated capital investment and is usually calculated at the end of each year.
- **Salvage value:** Estimated trade-in or market value at the end of the asset's useful life. It may be positive, negative, or zero. It can be expressed as a dollar amount or as a percentage of the first cost.
- **Market value:** This is the estimated amount realizable if the asset were sold in an open market. This amount may be different from the book value.
- **Recovery period:** This is the depreciable life of an asset in years. Often there are different n values for book and tax depreciations. Both values may be different from the asset's estimated productive life.
- **Depreciation or recovery rate:** This is the fraction of the first cost removed by depreciation each year. Depending on the method of depreciation, this rate may be different for each recovery period.
- **Half-year convention:** This is used with the Modified Accelerated Cost Recovery System (MACRS) depreciation method, which will be discussed later. It assumes that assets are placed in service or disposed of in midyear, regardless of when during the year these placements or disposal actually occur. There are also midquarter and midmonth conventions.

6.2   Depreciation methods

There are five principal methods of depreciation:

- Classical (historical) depreciation methods:
  - Straight-Line (SL)
  - Declining balance (DB)
  - Sum-of-years digits (SYD)
- Unit-of-production
- Modified accelerated cost recovery system (MACRS)

In describing each of these methods, let

n = recovery period in years
B = first cost, unadjusted basis, or basis
S = estimated salvage value
D_t = annual depreciable charge
MV = market value
BV_t = book value after period t
d = depreciation rate = $1/n$
t = year (t = 1, 2, 3, ..., n)

6.2.1   Straight-line (SL) method

This is the simplest and best-known method of depreciation. It assumes that a constant amount is depreciated each year over the depreciable (useful) life of the asset; hence, the book value decreases linearly with time. The SL method is considered the standard against which other depreciation models are compared. It offers an excellent representation of an asset used regularly over an estimated period, especially for book depreciation purposes. The annual depreciation charge is

$$D_t = \frac{B-S}{n} = (B-S)d \qquad (6.1)$$

The BV after t year is

$$BV_t = B - \frac{t}{n}(B-S) = B - tD_t \qquad (6.2)$$

Example 6.1

A piece of new petroleum-drilling equipment has a cost basis of $5000 and a 6-year depreciable life. The estimated salvage value of the machine is $500 at the end of 6 years. What are the annual depreciation amounts and the book value at the end of each year using the SL method?

Solution

For this problem, the basis (B) = $5000, S = $500, and n = 6. To compute the annual depreciation amounts and the book value, we use Equation 6.1 and Equation 6.2, respectively. Depreciation analysis is another module in ENGINEA. Using ENGINEA, we obtain the following results:

To use this module, we select the depreciation method, input the initial cost (basis), the salvage value, and the life (depreciable life). We click on "Calculate" to view the depreciation table.

As expected, the annual depreciation amount is constant at $750. In addition, the BV at the end of year 6 is $500, which is equal to the salvage value at the end of the depreciable life.

6.2.2　Declining balance (DB) method

This method is commonly applied as the book depreciation method in the industry because it accelerates the write-off of an asset value. It is also called the *Fixed (Uniform) Percentage* method; therefore, a constant depreciation rate is applied to the book value of the asset. According to the Tax Reform Act of 1986, two rates are applied to the straight-line rate: 150% and 200%. If 150% is used, it is called the DB method, and if 200% is used, it is called the Double Declining Balance (DDB) method. The DB annual depreciation charge is

$$D_t = \frac{1.5B}{n}\left(1 - \frac{1.5B}{n}\right)^{t-1} \tag{6.3}$$

The total DB depreciation at the end of t years is

$$B\left[1-\left(1-\frac{1.5}{n}\right)^{t}\right]=B\left[1-\left(1-d\right)^{t}\right] \qquad (6.4)$$

The BV at the end of t years is

$$B\left(1-\frac{1.5}{n}\right)^{t}=B\left(1-d\right)^{t} \qquad (6.5)$$

For the DDB (200% depreciation) method, substitute 2.0 for 1.5 in Equation 6.3, Equation 6.4, and Equation 6.5.

It should be noted that salvage value is not used in the equations for the DB or DDB methods; therefore, these methods are independent of the salvage value of the asset. The implication of this is that the depreciation schedule may go below an implied salvage value, above an implied salvage value, or just at the implied salvage value. Any of these three situations is possible in the real world. However, the U.S. Internal Revenue Service (IRS) does not permit a deduction for depreciation charges below the salvage value, whereas companies generally do not like to deduct depreciation charges that would keep the book value above the salvage value. The solution to this problem is to use a composite depreciation method. The IRS provides that a taxpayer may switch from DB or DDB to SL at any time during the life of an asset. However, the question is when it is a better time to switch. The criterion used to determine when to switch is based on the maximization of the present value of the total depreciation. Figure 6.1 shows a graphical depiction of the three possible profiles of salvage values (S) in relation to ending book value (BV). Part (a) of the figure shows ending book value being below the salvage value, part (b) shows ending book value being

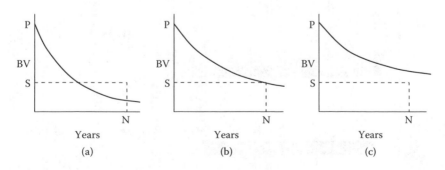

Figure 6.1 Effect of salvage value on DDB depreciation.

equal to the salvage value, and part (c) shows the ending book value being above the salvage value. Part (b) is the desirable scenario. If Part (a) occurs, less depreciation should be charged in the later years, or depreciation should be terminated to bring the BV up to the estimated salvage value at the end of N years. If Part (c) occurs, more depreciation should be charged in the later years to bring the BV down to the estimated salvage value at the end of N years. So, we have the following options:

Resolution of (a): Termination of depreciation as soon as value is reached.
Resolution of (c): Switch from a *declining balance* to *straight-line* depreciation in any year when the latter yields a larger depreciation charge.

Example 6.2

Repeat Example 6.1 using the DB and DDB methods.

Solution

Again, we use ENGINEA to solve this problem. The annual depreciation and BV at the end of each year are obtained by substituting the given values into Equation 6.3 and Equation 6.5, respectively.

For DB using 150%:

Depreciation Analysis 1

Depreciation Methods

- ○ Straight Line (SL)   ○ Double Declining Balance (DDB)
- ○ Sum of Years Digits (SOYD)   ○ Modified Accelerated Cost Recovery System (MACRS)
- ◉ Declining Balance (DB)   [1.5]

Initial Cost   [5000]

Salvage Value   [500]

Life   [6]

Period	Depreciation	Book Value
0	0.00	5000.00
1	1250.00	3750.00
2	937.50	2812.50
3	703.13	2109.38
4	527.34	1582.03
5	395.51	1186.52
6	296.63	889.89

[Calculate]

For DDB:

6.2.3   *Sums-of-years' digits (SYD) method*

This method results in larger depreciation charges during the early years of an asset (than SL) and smaller charges during the latter part of the estimated useful life; however, write-off is not as rapid as for DDB or MACRS (modified accelerated cost recovery system). As in the case of the SL method, this method uses the salvage value in computing the annual depreciation charge. The annual depreciation charge is as follows:

$$D_t = \frac{n-t+1}{SUM}(B-S) = d_t(B-S)$$

$$SUM = \frac{n(n+1)}{2}$$

(6.6)

The BV at the end of t years is

$$BV_t = B - \frac{t\left(n - \dfrac{t}{2} + 0.5\right)}{SUM}(B-S)$$

(6.7)

Example 6.3

Repeat Example 6.1 using the SYD method.

Solution

Equation 6.6 and Equation 6.7 can be used to obtain the annual depreciation and BV at the end of each year. The result using ENGINEA is displayed here:

6.2.4   *Modified accelerated cost recovery system (MACRS) method*

This is the only approved tax depreciation method in the U.S. It is a composite method that automatically switches from DB or DDB to SL depreciation. The switch usually takes place whenever the SL depreciation results in larger depreciation charges, that is, a more rapid reduction in the BV of the asset. One advantage of the MACRS method is that it assumes that the salvage value is zero; therefore, it always depreciates to zero. Another outstanding advantage of this method is that it uses property classes that specify the recovery periods, n. The method adopts the half-year convention, which makes the actual recovery period 1 year longer than the specified period. The half-year convention means that the IRS assumes that the assets are placed in service halfway through the year, no matter when the assets were actually placed in service. This convention is also applicable when the asset is disposed of before the end of the depreciation period. The MACRS method consists of two systems for computing depreciation deductions: the General Depreciation System (GDS) and the Alternative Depreciation System (ADS). The ADS system is used for properties placed in any tax-exempt use, as well as properties used predominantly outside the U.S. The system provides a longer recovery period and uses only the SL method of depreciation. Therefore, this system is generally not considered an option for economic analysis. However, any property that qualifies for GDS can be depreciated under ADS, if preferred.

The following information is required to depreciate an asset using the MACRS method:

- The cost basis
- The date the property was placed in service
- The property class and recovery period
- The MACRS depreciation system to be used (GDS or ADS)
- The time convention that applies (for example, the half-year or quarter-year convention)

The steps involved in using the MACRS depreciation method follow:

1. Determine the property class of the asset being depreciated using published tables (see Table 6.1 and Table 6.2). Any asset not in any of the stated classes is automatically assigned a 7-year recovery period under the GDS system.
2. After the property class is known, find the appropriate published depreciation schedule (see Table 6.3). For nonresidential real property, see Table 6.4.
3. The last step is to multiply the asset's cost basis by the depreciation schedule for each year to get the annual depreciation charge as stated in Equation 6.8.

The MACRS annual depreciation amount is

$$D_t = \left(\text{First cost}\right) \times \left(\text{Tabulated depreciation schedule}\right)$$

$$= d_t B$$

(6.8)

The annual book value is

$$BV_t = \text{First cost} - \text{Sum of accumulated depreciation}$$

$$= B - \sum_{j=1}^{t} D_j$$

(6.9)

Table 6.1 Property Classes Based on Asset Description

Asset Class	Asset Description	Class Life (Years) ADR[a]	MACRS Property Class (Years) GDS	ADS
00.11	Office furniture, fixtures, and equipment	10	7	10
00.12	Information systems: computers/ peripherals	6	5	6
00.22	Automobiles, taxis	3	5	6
00.241	Light general-purpose trucks	4	5	6
00.25	Railroad cars and locomotives	15	7	15
00.40	Industrial steam and electric distribution	22	15	22
01.11	Cotton gin assets	10	7	10
01.21	Cattle, breeding or dairy	7	5	7
13.00	Offshore drilling assets	7.5	5	7.5
13.30	Petroleum refining assets	16	10	16
15.00	Construction assets	6	5	6
20.1	Manufacture of motor vehicles	12	7	12
20.10	Manufacture of grain and grain mill products	17	10	17
20.20	Manufacture of yarn, thread, and woven fabric	11	7	11
24.10	Cutting of timber	6	5	6
32.20	Manufacture of cement	20	15	20
48.10	Telephone distribution plant	24	15	24
48.2	Radio and television broadcasting equipment	6	5	6
49.12	Electric utility nuclear production plant	20	15	20
49.13	Electric utility steam production plant	28	20	28
49.23	Natural gas production plant	14	7	14
50.00	Municipal wastewater treatment plant	24	15	24
80.00	Theme and amusement assets	12.5	7	12.5

[a] ADR = Asset Depreciation Range.

Example 6.4

Repeat Example 6.1 using MACRS.

Solution

Because this is petroleum-drilling equipment, the property class is 5 years, according to Table 6.2. Therefore, the applicable percentage is as tabulated in the third column of Table 6.3. These tabulated values are substituted into Equation 6.8 to compute the annual depreciation amount. The sum of the accumulated depreciation is substituted into Equation 6.9 to compute the BV at the end of each year. The depreciation schedule using ENGINEA is as follows:

Because of the assumption of the half-life convention, the equipment is depreciated for 6 years even though the property class is 5 years. The BV at the end of the 6th year is zero because MACRS assumes that the salvage value is always zero.

Practice problems: Depreciation methods

6.1 A construction company purchased a piece of equipment for $220,000 five years ago. The current market value of the equipment is $70,000, and its book value is $50,000. A new piece of similar equipment is currently advertised for $350,000. What are the annual depreciation amounts at end of each year using the straight-line method?

6.2 A manufacturing company has bought some machinery for the manufacture of electronic devices with a cost basis of $2 million. Its market value at the end of 6 years is estimated to be $200,000. The before-tax MARR is 20% per year. Develop a GDS depreciation schedule for this machinery.

6.3 Some front-office furniture cost $85,000 and has an estimated salvage value of $10,000 at the end of 7 years of useful life. Compute the depreciation schedules and BV to the end of the useful life of the furniture by the following methods:
   a. SL
   b. SYD
   c. DDB
   d. DB
   e. MACRS

Table 6.2 MACRS GDS Property Classes[a]

Property Class	

Personal Property (All Property Except Real Estate)

3-Year property	Special handling devices for food processing and beverage manufacture
	Special tools for the manufacture of finished plastic products, fabricated metal products, and motor vehicles
	Property with ADR class life of 4 years or less
5-Year property	Automobiles[a] and trucks
	Aircraft (of non-air-transport companies)
	Equipment used in research and experimentation
	Computers
	Petroleum drilling equipment
	Property with ADR class life of more than 4 years and less than 10 years
7-Year property	All other property not assigned to another class
	Office furniture, fixtures, and equipment
	Property with ADR class life of 10 years or more and less than 16 years
10-Year property	Assets used in petroleum refining and certain food products
	Vessels and water transportation equipment
	Property with ADR class life of 16 years or more and less than 20 years
15-Year property	Telephone distribution plants
	Municipal sewage treatment plants
	Property with ADR class life of 20 years or more and less than 25 years
20-Year property	Municipal sewers
	Property with ADR class life of 20 years or more and less than 25 years

Real Property (Real Estate)

27.5 Years	Residential rental property (does not include hotels and motels)
39 Years	Nonresidential real property

[a] The depreciation deduction for automobiles is limited to $7660 (maximum) the first tax year, $4900 the second year, $2950 the third year, and $1775 per year in subsequent years.

Source: U.S. Department of the Treasury (2007), Internal Revenue Service Publication 946, *How to Depreciate Property*. Washington, D.C.

6.4 Make a plot of the depreciation schedules computed in Problem 6.3 and discuss the depreciation implication of each schedule for small-size, medium-size, and large-size companies.

6.5 The MACRS method is a hybrid depreciation method of either DB or DDB with a switch to the SL method. Using the tabulated percentage in Table 6.3, determine which property classes use DB with a switch to SL, and which classes use DDB with a switch to SL.

Table 6.3 MACRS Depreciation for Personal Property Based on the Half-Year Convention

Recovery Year	Applicable Percentage for Property Class					
	3-Year Property	5-Year Property	7-Year Property	10-Year Property	15-Year Property	20-Year Property
1	33.33	20.00	14.29	10.00	5.00	3.750
2	44.45	32.00	24.49	18.00	9.50	7.219
3	14.81	19.20	17.49	14.40	8.55	6.677
4	7.41	11.52	12.49	11.52	7.70	6.177
5		11.52	8.93	9.22	6.93	5.713
6		5.76	8.92	7.37	6.23	5.285
7			8.93	6.55	5.90	4.888
8			4.46	6.55	5.90	4.522
9				6.56	5.91	4.462
10				6.55	5.90	4.461
11				3.28	5.91	4.462
12					5.90	4.461
13					5.91	4.462
14					5.90	4.461
15					5.91	4.462
16					2.95	4.461
17						4.462
18						4.461
19						4.462
20						4.461
21						2.231

Table 6.4 MACRS Depreciation for Real Property (Real Estate)

Recovery Year	Recovery Percentage for Nonresidential Real Property (Month Placed in Service)											
	1	2	3	4	5	6	7	8	9	10	11	12
1	2.461	2.247	2.033	1.819	1.605	1.391	1.177	0.963	0.749	0.535	0.321	0.107
2–39	2.564	2.564	2.564	2.564	2.564	2.564	2.564	2.564	2.564	2.564	2.564	2.564
40	0.107	0.321	0.535	0.749	0.963	1.177	1.391	1.605	1.819	2.033	2.247	2.461

chapter seven

Break-even analysis

The term *break-even analysis* refers to an analysis to determine the point at which there exists a balanced performance level when a project's income is equal to its expenditure. The total cost of an operation is expressed as the sum of the fixed and variable costs with respect to output quantity. That is,

$$TC(x) = FC + VC(x)$$

where x is the number of units produced, $TC(x)$ is the total cost of producing x units, FC is the total fixed cost, and $VC(x)$ is the total variable cost associated with producing x units. The total revenue (TR) resulting from the sale of x units is defined as follows:

$$TR(x) - px$$

where p is the price per unit. The profit (P) due to the production and sale of x units of the product is calculated as

$$P(x) = TR(x) - TC(x)$$

The *break-even point* of an operation is defined as the value of a given parameter that will result in neither profit nor loss. The parameter of interest may be the number of units produced, the number of hours of operation, the number of units of a resource type allocated, or any other measure of interest. At the break-even point, we have the following relationship:

$$TR(x) = TC(x) \text{ or } P(x) = 0$$

7.1 Illustrative examples

In some cases, there may be a known mathematical relationship between cost and the parameter of interest. For example, there may be a linear cost relationship between the total cost of a project and the number of units

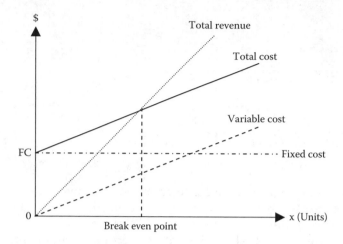

Figure 7.1 Break-even point for a single project.

produced. Such cost expressions facilitate straightforward break-even analysis. Figure 7.1 shows an example of a break-even point for a single project. Figure 7.3 shows examples of multiple break-even points that exist when multiple projects are compared. When two project alternatives are compared, the break-even point refers to the point of indifference between the two alternatives, or the point at which it does not matter which alternative is chosen. In Figure 7.3, x_1 represents the point where projects A and B are equally desirable, x_2 represents the point at which A and C are equally desirable, and x_3 represents the point at which B and C are equally desirable. The figure shows that if we are operating below a production level of x_2

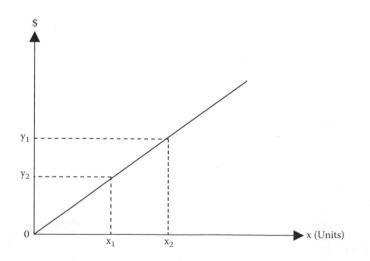

Figure 7.2 Incremental production mapped to incremental cost.

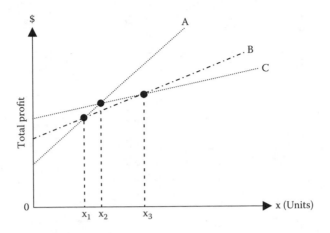

Figure 7.3 Break-even points for multiple projects.

units, then project C is the preferred project among the three. If we are operating at a level more than x_2 units, then project A is the best choice.

For $x_2 - x_1$ incremental production, there is $y_2 - y_1$ incremental cost.

Example 7.1

Three project alternatives are being considered for producing a new product. The required analysis will determine which alternative should be selected on the basis of how many units of the product are produced per year. Based on past records, there is a known relationship between the number of units produced per year, x, and the net annual profit, $P(x)$, from each alternative. The level of production is expected to be between 0 and 250 units per year. The net annual profits (in thousands of dollars) are as follows for each alternative:

Project A: $P(x) = 3x - 200$
Project B: $P(x) = x$
Project C: $P(x) = (1/50)x^2 - 300$

This problem can be solved mathematically by finding the intersection points of the profit functions and evaluating the respective profits over the given range of product units. However, it can also be solved by a graphical approach. Figure 7.4 shows the break-even chart, which is simply a plot of the profit functions. The plot shows that Project B should be selected if between 0 and 100 units are to be produced. Project A should be selected if between 100 and 178.1 units (178 physical units) are to be produced. Project C should be selected if more than 178 units are to be produced. It should be noted that if less than 66.7 units (66 physical units) are produced, Project A will generate a net loss rather than a net profit. Similarly, Project C will generate losses if less than 122.5 units (122 physical units) are produced.

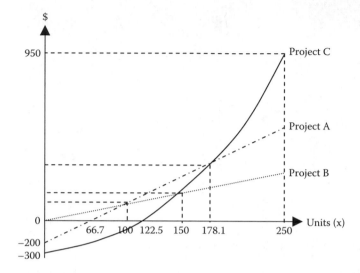

Figure 7.4 Plot of profit functions.

Example 7.2

RAB General Hospital must decide whether it should perform some medical tests for its patients in-house or whether it should pay a specialized laboratory to undertake the procedures. To perform the procedures in-house, the hospital will have to purchase computers, printers, and other peripherals at a cost of $15,000. The equipment will have a useful life of 3 years, after which it will be sold for $3,000. The employee who manages the tests will be paid $48,000 per year. In addition, each test will have an average cost of $4. Alternatively, the company can outsource the procedure at a flat-rate fee of $20 per test. At an interest rate of 10% per year, how many tests must the hospital perform each year in order for the alternatives to break even?

Solution

Let X = number of tests per year. The required equation is

$$I(A/P, i\%, n) - \text{Annual Salary} + SV(A/F, i\%, n) - 4X = -20X$$

where I is the cost of purchasing the equipment, and SV is its salvage value after 3 years. The left-hand side of the equation shows the cost when the procedures are performed in-house, and the right-hand side of the equation represents the cost when the procedures are performed by the specialized laboratory.

Substituting values, the equation becomes:

$$-15,000\left(A/P,10\%,3\right)-48,000+3,000\left(A/F,10\%,3\right)-4X=-20X$$

$$-15,000\left(0.40211\right)-48,000+3,000\left(0.30211\right)=-20X+4X$$

$$-53,125.32=-16X$$

$$\therefore X\approx3,320$$

The preceding equation shows the break-even number of tests to be 3,320. If the hospital plans to perform less than 3,320 tests in a year (that is, $X < 3,320$), it will be better to outsource the procedures; but if $X > 3,320$, it will be better to perform the procedures in-house. When $X = 3,320$, it does not matter whether the procedure is performed in-house or outsourced.

7.2 Profit ratio analysis

Break-even charts offer the opportunity for several different types of analysis. In addition to the break-even points, other measures of worth, or criterion measures, may be derived from the charts. For example, a measure called the *profit ratio* permits one to obtain a further comparative basis for competing projects. The *profit ratio* is defined as the ratio of the profit area to the sum of the profit-and-loss areas in a break-even chart. That is,

$$\text{Profit ratio} = \frac{\text{Area of profit region}}{\text{Area of profit region} + \text{Area of loss region}}$$

Example 7.3

Suppose that the expected revenue and the expected total costs associated with a project are given, respectively, by the following expressions:

$$R(x) = 100 + 10x$$

$$TC(x) = 2.5x + 250$$

where x is the number of units produced and sold from the project. In Figure 7.5, we can see that the break-even chart shows the break-even point to be 20 units. Net profits will be realized from the project if more than 20 units are produced, and net losses will result if fewer than 20 units are produced. It should be noted that the revenue function in Figure 7.5 represents an unusual case in which a revenue of $100 is realized even when zero units are produced.

Suppose it is desired to calculate the profit ratio for this project if the number of units that can be produced is limited to between 0 and 100 units.

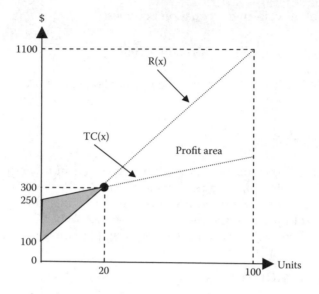

Figure 7.5 Area of profit vs. area of loss.

From Figure 7.5, the surface areas of the profit region and the loss region can both be calculated by using the standard formula for finding the area of a triangle, Area = (1/2)(Base)(Height). Using this formula, we have the following:

$$\text{Area of profit region} = \frac{1}{2}(\text{Base})(\text{Height})$$

$$= \frac{1}{2}(1100 - 500)(100 - 20)$$

$$= 24,000 \text{ square units}$$

$$\text{Area of loss region} = \frac{1}{2}(\text{Base})(\text{Height})$$

$$= \frac{1}{2}(250 - 100)(20)$$

$$= 1500 \text{ square units}$$

Thus, the profit ratio is computed as

$$\text{Profit Ratio} = \frac{24000}{24000 + 1500}$$

$$= 0.9411$$

$$= 94.11\%$$

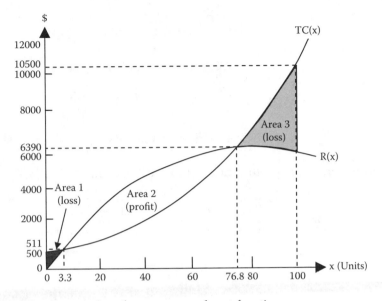

Figure 7.6 Break-even chart for revenue and cost functions.

The profit ratio may be used as a criterion for selecting among project alternatives. Note that the profit ratios for all the alternatives must be calculated over the same values of the independent variable. The project with the highest profit ratio will be selected as the desired project. For example, Figure 7.6 presents the break-even chart for an alternate project, say Project II. It is seen that both the revenue and cost functions for the project are non-linear. The revenue and cost are defined as follows:

$$R(x) = 160x - x^2$$

$$TC(x) = 500 + x^2$$

If the cost and/or revenue functions for a project are not linear, the areas bounded by the functions may not be easy to determine. For those cases, it may be necessary to use techniques such as definite integrals to find the areas. Figure 7.6 indicates that the project generates a loss if less than 3.3 units (3 actual units) or more than 76.8 units (76 actual units) are produced. The respective profit and loss areas on the chart are calculated as follows:

$$\text{Area 1 (loss)} = \int_0^{3.3} \left[\left(500 + x^2\right) - \left(160x - x^2\right) \right] dx$$

$$= 802.8 \text{ unit-dollars}$$

$$\text{Area 2 (profit)} = \int_{3.3}^{76.8} \left[\left(160x - x^2 \right) - \left(500 + x^2 \right) \right] dx$$

$$= 132,272.08 \text{ unit-dollars}$$

$$\text{Area 3 (loss)} = \int_{76.8}^{100} \left[\left(500 + x^2 \right) - \left(160x - x^2 \right) \right] dx$$

$$= 48,135.98 \text{ unit-dollars}$$

Consequently, the profit ratio for Project II is computed as

$$\text{Profit ratio} = \frac{\text{Total area of profit region}}{\text{Total area of profit region} + \text{Total area of loss region}}$$

$$= \frac{132,272.08}{802.76 + 132,272.08 + 48,135.98}$$

$$= 72.99\%$$

The profit ratio approach evaluates the performance of each alternative over a specified range of operating levels. Most of the existing evaluation methods use single-point analysis with the assumption that the operating condition is fixed at a given production level. The profit ratio measure allows an analyst to evaluate the net yield of an alternative, given that the production level may shift from one level to another. It is possible, for example, for an alternative to operate at a loss for most of its early life but for it to generate large incomes to offset the earlier losses in its later stages. Conventional methods cannot easily capture this type of transition from one performance level to another. In addition to being used to compare alternate projects, the profit ratio may also be used for evaluating the economic feasibility of a single project. In such a case, a decision rule may be developed. An example of such a decision rule follows:

If a profit ratio is greater than 75%, accept the project.
If a profit ratio is less than or equal to 75%, reject the project.

Practice problems: Break-even analysis

7.1   Considering Example 7.2, if the RAB General Hospital anticipates that the number of tests to be conducted in a year varies between 500 and 5000, which option should be selected (in-house or out-source) based on the profit ratio analysis?

7.2   The capacity of a gear-producing plant is 4,600 gears per month. The fixed cost is $704,000 per month. The variable cost is $130 per gear, and the selling price is $158 per gear. Assuming that all products produced are sold, what is the break-even point in number of gears per month? What percent increase in the break-even point will occur if the fixed costs are increased by 20% and unit variable costs are increased by 4%?

7.3 A manufacturing firm produces and sells three different types of tires — car tires, van tires, and bus tires. Due to warehouse space constraints, the firm's production is limited to 20,000 car tires, 16,000 van tires, and 24,000 bus tires. The variable and fixed costs associated with each type of tire are as follows:

Tire	Cost Type	
Type	Variable Cost	Fixed Cost
Car	$17.00	$800,000/month
Van	$25.00	$1,200,000/month
Bus	$32.00	$1,600,000/month

If the selling price of a car, van, and bus tire is $100, $150, and $200 each, respectively, determine the break-even point for each type of tire.

7.4 A company is considering establishing a warehouse in either the southern or eastern part of the city where there is a potential market for their products. The management of the company needs to decide where to set up the warehouse by comparing the expenses that will be incurred if the warehouse is set up in either of the locations. Given the following data, compare the two sites in terms of their fixed, variable, and total costs. Which is the better location?

Cost Factor	South	East
Cost of land	$1.2 million	$1.4 million
Estimated cost of building materials	$600,000	$450,000
Number of laborers required	80	60
Hourly payment of laborers	$12/h	$10/h
Estimated miscellaneous expenses	$50,000	$62,000

7.5 A company produces a brand of sugar that is used for baking by several retail confectioneries. The fixed cost is $13,000 per month, and the variable cost is $4 per unit. The selling price per unit $p = $45 - 0.01D$. For this scenario,

a. Determine the optimal volume of this brand of sugar and confirm that a profit occurs instead of a loss at this demand.

b. Find the volume at which break-even occurs, that is, what is the domain of profitable demand?

chapter eight

Effects of inflation and taxes

The effect of inflation is an important consideration in financial and economic analysis of projects. *Inflation* can be defined as a decline in the purchasing power of money, and multiyear projects are particularly subject to its effects. Some of the most common causes of inflation include the following:

- An increase in the amount of currency in circulation
- A shortage of consumer goods
- An escalation of the cost of production
- An arbitrary increase of prices by resellers

The general effects of inflation are felt as an increase in the price of goods and a decrease in the worth of currency. In cash-flow analysis, return on investment (ROI) for a project will be affected by the time value of money as well as by inflation. The *real interest rate* (d) is defined as the desired rate of return in the absence of inflation. When we talk of "today's dollars" or "constant dollars," we are referring to the use of the real interest rate. The *combined interest rate* (i) is the rate of return combining the real interest rate and the rate of inflation. If we denote the *inflation rate* as j, then the relationship between the different rates can be expressed as follows:

$$1 + i = (1 + d)(1 + j)$$

Thus, the combined interest rate can be expressed as

$$i = d + j + dj$$

Note that if $j = 0$ (i.e., there is no inflation), then $i = d$. We can also define the *commodity escalation rate* (g) as the rate at which individual commodity prices escalate. This may be greater or less than the overall inflation rate.

In practice, several measures are used to convey inflationary effects. Some of these are the *Consumer Price Index*, the *Producer Price Index*, and the *Wholesale Price Index*. A *"market basket"* rate is defined as the estimate of inflation

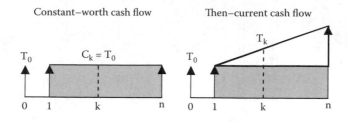

Figure 8.1 Cash flows for effects of inflation.

based on a weighted average of the annual rates of change in the costs of a wide range of representative commodities. A *"then-current" cash flow* is a cash flow that explicitly incorporates the impact of inflation. A *"con-stant-worth" cash flow* is a cash flow that does not incorporate the effect of inflation. The *real interest rate, d,* is used for analyzing constant-worth cash flows. Figure 8.1 shows constant-worth and then-current cash flows.

The then-current cash flow in the figure is the equivalent cash flow considering the effect of inflation. C_k is what it would take to buy a certain "basket" of goods after k time periods if there was no inflation. T_k is what it would take to buy the same "basket" in k time period if inflation were taken into account. For the constant worth cash flow, we have

$$C_k = T_0, k = 1, 2, ..., n$$

and for the then-current cash flow, we have

$$T_k = T_0(1 + j)^k, k = 1, 2, ..., n$$

where j is the inflation rate. If $C_k = T_0 = \$100$ under the constant worth cash flow, then we mean \$100 worth of buying power. If we are using the commodity escalation rate, g, then we obtain

$$T_k = T_0(1 + g)^k, k = 1, 2, ..., n$$

Thus, a then-current cash flow may increase based on both a regular inflation rate (j) and a commodity escalation rate (g). We can convert a then-current cash flow to a constant-worth cash flow by using the following relationship:

$$C_k = T_k(1 + j)^{-k}, k = 1, 2, ..., n$$

If we substitute T_k from the commodity escalation cash flow into the preceding expression for C_k, we get the following:

$$C_k = T_k \left(1+j\right)^{-k}$$

$$= T \left(1+g\right)^k \left(1+j\right)^{-k}$$

$$= T_0 \left[\left(1+g\right)/\left(1+j\right)\right]^k, \quad k = 1, 2, \ldots, n$$

Note that if $g = 0$ and $j = 0$, then $C_k = T_0$. That is, there is no inflationary effect. We now define effective commodity escalation rate (v) as

$$v = [(1 + g)/(1 + j)] - 1$$

and we can express the commodity escalation rate (g) as

$$g = v + j + vj$$

Inflation can have a significant impact on the financial and economic aspects of a project. In economic terms, inflation can be seen as the increase in the amount of currency in circulation such that there results a relatively high and sudden fall in its value. To a producer, inflation means a sudden increase in the cost of items that serve as inputs for the production process (equipment, labor, materials, etc). To the retailer, inflation implies an imposed higher cost of finished products. To an ordinary citizen, inflation portends an unbearable escalation of prices of consumer goods. All these aspects of inflation are interrelated in a project-management environment.

The amount of money supply, as a measure of a country's wealth, is controlled by that country's government. For various reasons, governments often feel impelled to create more money or credit to take care of old debts and pay for social programs. When money is generated at a faster rate than the growth of goods and services, it becomes a surplus commodity, and its value (purchasing power) will fall. This means that there will be too much money available to buy only a few goods and services. When the purchasing power of a currency falls, each individual in a product's life cycle has to dispense more of the currency in order to obtain the product. Some of the classic concepts of inflation are discussed here.

1. Increases in producers' costs are passed on to consumers. At each stage of a product's journey from producer to consumer, prices are escalated disproportionately in order to make a good profit. The overall increase in the product's price is directly proportional to the number of intermediaries it encounters on its way to the consumer. This type of inflation is called *cost-driven* (or *cost-push*) *inflation*.
2. Excessive spending power of consumers forces an upward trend in prices. This high spending power is usually achieved at the expense

of savings. The law of supply and demand dictates that the higher the demand, the higher the price. This type of inflation is known as *demand-driven* (or *demand-pull*) *inflation*.

3. The impact of international economic forces can induce inflation in a local economy. Trade imbalances and fluctuations in currency values are notable examples of international inflationary factors.

4. Increasing base wages of workers generate more disposable income and, hence, higher demands for goods and services. The high demand, consequently, creates a pull on prices. Coupled with this, employers pass on the additional wage cost to consumers through higher prices. This type of inflation is, perhaps, the most difficult to solve because union-set wages and producer-set prices almost never fall, at least not permanently. This type of inflation may be referred to as *wage-driven* (or *wage-push*) *inflation*.

5. Easy availability of credit leads consumers to "buy now and pay later" and, thereby, creates another loophole for inflation. This is a dangerous type of inflation because the credit not only pushes prices up but also leaves consumers with less money later on to pay for the credit. Eventually, many credits become uncollectible debts, which may then drive the economy into recession.

6. Deficit spending results in an increase in money supply and, thereby, creates less room for each dollar to get around. The popular saying, "a dollar doesn't go as far as it used to," simply refers to inflation in layman's terms. The different levels of inflation may be categorized in the following way.

8.1   Mild inflation

When inflation is mild (2 to 4%), the economy actually prospers. Producers strive to produce at full capacity in order to take advantage of the high prices to the consumer. Private investments tend to be brisk, and more jobs become available. However, the good fortune may only be temporary. Prompted by the prevailing success, employers become tempted to seek larger profits, and workers begin to ask for higher wages. They cite their employers' prosperous business as a reason to bargain for bigger shares of the business profit. So, we end up with a vicious cycle in which the producer asks for higher prices, the unions ask for higher wages, higher wages necessitate higher prices, and inflation starts an upward trend.

8.2   Moderate inflation

Moderate inflation occurs when prices increase at 5 to 9%. Consumers start purchasing more as an edge against inflation. They would rather spend their money now than watch it decline further in purchasing power. The increased market activity serves to fuel further inflation.

8.3   Severe inflation

Severe inflation is indicated by price escalations of 10% or more. *Double-digit inflation* implies that prices rise much faster than wages do. Debtors tend to be the ones who benefit from this level of inflation because they repay debts with money that is less valuable than the money that was borrowed.

8.4   Hyperinflation

When each price increase signals another increase in wages and costs, which again sends prices further up, the economy has reached a stage of malignant *galloping inflation* or hyperinflation. Rapid and uncontrollable inflation will destroy an economy. Currency becomes economically useless as the government prints it excessively to pay for obligations.

Inflation can affect any project in terms of raw materials procurement, salaries and wages, and/or difficulties with cost tracking. Some effects are immediate and easily observable, others subtle and pervasive. Whatever form it takes, inflation must be taken into account in long-term project planning and control. Large projects may be adversely affected by the effects of inflation in terms of cost overruns and poor resource utilization. The level of inflation will determine the severity of the impact on projects.

Example 8.1

An Internet company is considering investing in a new technology that would place its business above its competitors at the end of 3 years, when the project would be completed. Two of the company's top suppliers have been contacted to develop the technology. The yearly project costs of the new technology for each of the suppliers are as follows. If the Internet company uses an MARR of 15%, and if the general price inflation is assumed to be 2.5% per year over the next 3 years, which supplier should be given the contract?

Year	Supplier A (All values in Future Dollars)	Supplier B (All Values in Today's Dollars)
1	120,000	120,000
2	132,000	120,000
3	145,200	120,000

Solution

It must be noted that Supplier A presents all its costs in terms of future dollars, and Supplier B presents its in terms of today's dollars. To answer this question, we can either convert future dollars to today's dollars or convert today's dollars to future dollars. If the calculation is done properly, we should get the same results. From the question, $j = 0.025$ and $i = 0.15$.

Approach I: Convert today's dollars to future dollars and find the present value (*PV*) using *i* (the combined interest rate or MARR).

Using the equation Future dollar = Today's dollar $(1 + j)^k \equiv Tk = T_0(1 + j)^k$, the equivalent future dollars for Supplier B are

$$\text{Year 1: } 120,000 \times (1+0.025)^1 = 123,000$$

$$\text{Year 2: } 120,000 \times (1+0.025)^2 = 126,075$$

$$\text{Year 3: } 120,000 \times (1+0.025)^3 = 129,227$$

Finding the present value,

$$PV_A = 120,000(P/F,15\%,1) + 132,000(P/F,15\%, 2) + 145,200(P/F,15\%, 3)$$

$$= \$299,615$$

$$PV_B = 123,000(P/F,15\%,1) + 126,075(P/F,15\%, 2) + 129,227(P/F,15\%, 3)$$

$$= \$287,246$$

Therefore, Supplier B should be selected because its *PV* is the lesser of the two.

Approach II: Convert future dollars to today's dollars and find the *PV* using *d* (the real interest rate):

$$d = \frac{i-j}{1+j} = \frac{0.15 - 0.025}{1 + 0.025} = 12.2\%$$

Using the equation Today's dollar = Future dollar $(1 + j)^{-k} \equiv T_0 = T_k (1 + j)^{-k}$, the equivalent today's dollars for Supplier A are

$$\text{Year 1: } 120,000 \times (1+0.025)^{-1} = 117,073$$

$$\text{Year 2: } 132,000 \times (1+0.025)^{-2} = 125,640$$

$$\text{Year 3: } 145,200 \times (1+0.025)^{-3} = 134,833$$

Finding the present value,

$$PV_A = 117,073(P/F,12.2\%,1) + 125,640(P/F,12.2\%,2) + 134,833(P/F,12.2\%,3)$$

$$= \$299,605$$

$$PV_B = 120,000(P/A,12.2\%,3)$$

$$= \$287,232$$

Again, Supplier B should be selected because its *PV* is the lesser. The difference in *PV* is due to round-off errors.

8.5   Foreign-exchange rates

The idea of accounting for the effects of inflation on local investments can be extended to account for the effects of devalued currency on foreign investments. When local businesses invest in a foreign country, several factors come into consideration, such as when the initial investment is made and when the benefits are returned to the local business. As a result of changes in international businesses, exchange rates between currencies fluctuate, and countries continually devalue their currencies to satisfy international trade agreements.

Using the U.S. as the base country where the local business is situated, let

i_{us} = rate of return in terms of a market interest rate relative to U.S. dollars.
i_{fc} = rate of return in terms of a market interest rate relative to the currency of a foreign country.
f_e = annual devaluation rate between the currency of a foreign country and the U.S. dollar. A positive value means that the foreign currency is being devalued relative to the U.S. dollar. A negative value means that the U.S. dollar is being devalued relative to the foreign currency.

Therefore, the rate of return of the local business with respect to a foreign country can be given by the following:

$$i_{us} = \frac{i_{fc} - f_e}{1 + f_e}$$

(8.1)

$$i_{fc} = i_{us} + f_e + f_e(i_{us})$$

Example 8.2

A European company is considering investing in a technology in the U.S. The company has a MARR of 20% on investments in Europe. Additionally, its currency, the RAW, is very strong, with the U.S. dollars being devalued an average of 4% annually with respect to it.

    a.   Based on a before-tax analysis, would the European company want to invest in a technology proposal in the U.S. for $10 million that would repay $2.5 million each year for 10 years with no salvage value?

    b.   If a U.S. company with a MARR of 20% wanted to invest in this European country, what MARR should it expect on cash flows in RAWs?

    Solution

    a.   From the question, for the European company,

$$f_e = -4\% \text{ (because the U.S. dollar is being devalued)}$$

$$i_{fc} = 20\%$$

$$n = 10$$

$$i_{US} = \frac{i_{fc} - f_e}{1 + f_e} = \frac{0.2 - (-0.04)}{1 + (-0.04)} = 0.25$$

$$PV_{i_{US}} = 2,500,000(P/A, 25\%, 10) = \$8,926,258$$

Because the *PV* of the benefits of the investment is less than the amount invested, the European company would not want to invest in this proposal.

    b.   For the U.S. company,

$$i_{fc} = i_{US} + f_e + f_e(i_{US})$$

$$= 0.20 + (-0.04) + (-0.04)(0.20)$$

$$= 15.2\%$$

Therefore, the U.S. company should not expect a MARR of more than 15.2% in RAWs.

8.6   After-tax economic analysis

For complete and accurate economic analysis results, both the effects of inflation and taxes must be taken into consideration, especially when alternatives are being evaluated. Taxes are an inevitable burden such that their effects must be accounted for in economic analysis. There are several types of taxes:

- Income taxes: These are taxes assessed as a function of gross revenue less allowable deductions, and are levied by the federal government, and most state and municipal governments.
- Property taxes: They are assessed as a function of the value of property owned, such as land, buildings, and equipment. They are mostly levied by municipal, county, or state governments.
- Sales taxes: These are assessed on purchases of goods and services; hence, sales taxes are independent of gross income or profits. They are normally levied by state, municipal, or county governments. Sales taxes are relevant in economic analysis only to the extent that they add to the cost of items purchased.
- Excise taxes: These are federal taxes assessed as a function of the sale of certain goods or services often considered nonnecessities. They are usually charged to the manufacturer of the goods and services, but a portion of the cost is passed on to the purchaser.

Income taxes are the most significant type of tax encountered in economic analysis; therefore, the effects of income taxes can be accounted for using these relations. Let
TI = Taxable income (amount upon which taxes are based).
T = Tax rate (percentage of taxable income owed in taxes).
$NPAT$ = Net Profit after Taxes (taxable income less income taxes each year. This amount is returned to the company).

Therefore,

$$TI = \text{gross income} - \text{expenses} - \text{depreciation (depletion) deductions}$$

$$T = TI \times \text{Applicable tax rate} \tag{8.1}$$

$$NPAT = TI(1-T)$$

The tax rate used in economic analysis is usually the effective tax rate, and it is computed using this relationship:

$$\text{Effective tax rates } (Te) = \text{state rate} + (1 - \text{state rate})(\text{federal rate}) \tag{8.2}$$

Therefore,

$$T = (\text{Taxable Income})(Te) \tag{8.3}$$

8.7 Before-tax and after-tax cash flow

The only difference between a before-tax cash flow (BTCF) and an after-tax cash flow (ATCF) is that ATCF includes expenses (or savings) due to income taxes and uses an after-tax MARR to calculate equivalent worth. Hence, after-tax cash flow is the before-tax cash flow less taxes. The after-tax MARR is usually smaller than the before-tax MARR, and they are related by the following equation:

$$\text{After-tax MARR} \cong \left(\text{Before-tax MARR}\right)\left(1 - Te\right) \tag{8.4}$$

8.8 Effects of taxes on capital gain

Capital gain is the amount incurred when the selling price of a property exceeds its first cost. Because future capital gains are difficult to estimate, they are not detailed in after-tax study. However in actual tax law, there is no difference between a short-term or long-term gain.

Capital loss is the loss incurred when a depreciable asset is disposed of for less than its current book value. An economic analysis does not usually account for capital loss because it is not easily estimated for alternatives. However, after-tax replacement analysis should account for any capital loss. For economic analysis, this loss provides a tax saving in the year of replacement.

Depreciation recapture occurs when a depreciable asset is sold for more than its current book value. Therefore, depreciation recapture is the selling price less the book value. This is often present in after-tax analysis. When the MACRS system of depreciation is used, the estimated salvage value of an asset can be anticipated as the depreciation recapture because MACRS assumes zero salvage value.

Therefore, the taxable income equation can be rewritten as

$$TI = \text{gross income} - \text{expenses} - \text{depreciation (depletion) deductions}$$
$$+ \text{depreciation recapture} + \text{capital gain} - \text{capital loss} \tag{8.5}$$

8.9 After-tax computations

The ATCF estimates are used to compute the net present value (NPV) or net annual value (NAV) or net future value (NFV) at the after-tax MARR. The same logic applies as for the before-tax evaluation methods discussed in Chapter 5; however, the calculations required for after-tax computations are certainly more involved than those for before-tax analysis. The major elements in an after-tax economic analysis are these:

- Before-tax cash flow
- Depreciation
- Taxable income

- Income taxes
- After-tax cash flow

The incorporation of tax and depreciation aspects into the before-tax cash flow produces the after-tax cash flow, upon which all the computational procedures presented earlier can be applied to obtain measures of economic worth. The measures emanating from the after-tax cash flow are referred to as after-tax measures of economic worth.

Practice problems: Effects of inflation and taxes

8.1   You are considering buying some equipment to recover materials from an effluent at your chemical plant. These chemicals will have a market value of $157,000 next year and should increase in value at an inflation rate of 7.5% each year. Your company is using a planning period of 7 years for this type of project and a MARR of 15%. How much can you afford to pay for the recovery equipment? Ignore the effect of taxes.

8.2   On January 1, 1995, the National Price Index was 208.5, and on January 1, 2005, it was 516.71. What was the inflation rate compounded annually over that 10-year period?

8.3   A manufacturing machine cost $65,000 in 1990, and an equivalent model 5 years later cost $98,600. If inflation is considered the cause of the increase, what was the average annual rate of inflation?

8.4   A project has the following estimates in future dollars. If the inflation rate is 7%, what are the equivalent estimates in today's dollars?

EOY	Future Dollars	Today's Dollars
0	100,000	
1	22,300	
2	24,851	
3	27,680	
4	30,815	
5	34,290	
6	38,140	
7	42,404	
8	47,126	
9	52,354	
10	78,141	

8.5   (a) What is the rate of return of the project described in Problem 8.4 in terms of future dollars and in terms of today's dollars? Discuss the difference between these values.

(b) What is the discounted payback period of the project described in Problem 8.4 in terms of future dollars and in terms of today's dollars if the MARR is as computed in Problem 8.5 (a)?

chapter nine

Advanced cash-flow analysis techniques

In some cases, an analyst must resort to more advanced or intricate economic analysis in order to take into account the unique characteristics of a project. This chapter presents a collection of such techniques, which are based on an application of the computational processes presented in earlier chapters. Computation of amortization of capitals, tent cash-flow analysis, and equity break-even point calculation are presented in this chapter as additional techniques of economic analysis.

9.1 Amortization of capitals

Many capital investment projects are financed with external funds repaid according to an amortization schedule. A careful analysis must be conducted to ensure that the company involved can financially handle the amortization schedule, and here, a computer program such as GAMPS (Graphic Evaluation of Amortization Payments) might be useful for this purpose. Such a program analyzes installment payments, the unpaid balance, principal amount paid per period, total installment payment, and current cumulative equity. It also calculates the "equity break-even point" for the debt being analyzed. The equity break-even point indicates the time when the unpaid balance on a loan is equal to the cumulative equity on the loan. This is discussed in a later section of this chapter. With this or a similar program, the basic cost of servicing the project debt can be evaluated quickly. A part of the output of the program presents the percentage of the installment payment going into the equity and the interest charge, respectively. The computational procedure for analyzing project debt follows the steps outlined here:

1. Given a principal amount, P, a periodic interest rate, i (in decimals), and a discrete time span of n periods, the uniform series of equal end-of-period payments needed to amortize P is computed as

$$A = \frac{P\left[i(1+i)^n\right]}{(1+i)-1}$$

It is assumed that the loan is to be repaid in equal monthly payments. Thus, $A(t) = A$ for each period t throughout the life of the loan.

2. The unpaid balance after making t installment payments is given by

$$U(t) = \frac{A\left[1-(1+i)^{(t-n)}\right]}{i}$$

3. The amount of equity or principal amount paid with installment payment number t is given by

$$E(t) = A(1+i)^{t-n-1}$$

4. The amount of interest charge contained in installment payment number t is derived to be

$$I(t) = A\left[1-(1+i)^{t-n-1}\right]$$

where $A = E(t) + I(t)$.

5. The cumulative total payment made after t periods is denoted by

$$C(t) = \sum_{k=1}^{t} A(k)$$

$$= \sum_{k=1}^{t} A$$

$$= (A)(t)$$

6. The cumulative interest payment after t periods is given by

$$Q(t) = \sum_{x=1}^{t} I(x)$$

7. The cumulative principal payment after t periods is computed as

$$S(t) = \sum_{k=1}^{t} E(k)$$

$$= A \sum_{k=1}^{t} (1+i)^{-(n-k+1)}$$

$$= A \left[\frac{(1+i)^t - 1}{i(1+i)^n} \right]$$

where

$$\sum_{n=1}^{t} x^n = \frac{x^{x+1} - x}{x-1}$$

8. The percentage of interest charge contained in installment payment number t is

$$f(t) = \frac{I(t)}{A}(100\%)$$

9. The percentage of cumulative interest charge contained in the cumulative total payment up to and including payment number t is

$$F(t) = \frac{Q(t)}{C(t)}(100\%)$$

10. The percentage of cumulative principal payment contained in the cumulative total payment up to and including payment number t is

$$H(t) = \frac{S(t)}{C(t)}$$

$$= \frac{C(t) - Q(t)}{C(t)}$$

$$= 1 - \frac{Q(t)}{C(t)}$$

$$= 1 - F(t)$$

Table 9.1 Amortization Schedule for Financed Project

ENGINEA: Engineering Economic Analysis - [Loan / Mortgage Analysis 1]

File   Analysis   Factor Table   Window   Help

Loan Amount: 500000
Annual Nominal Interest Rate: 10   %
Length: 15   years

Calculate   Graph

t	U(t)	A(t)	E(t)	I(t)	C(t)	Q(t)	S(t)	I(t)	F(t)	H(t)
1	498793.64	5373.03	1206.36	4166.67	5373.03	4166.67	1206.36	77.55	77.55	22.45
2	497577.23	5373.03	1216.41	4156.61	10746.05	8323.28	2422.77	77.36	77.45	22.64
3	496350.68	5373.03	1226.55	4146.48	16119.08	12469.76	3649.32	77.17	77.36	22.83
4	495113.91	5373.03	1236.77	4136.26	21492.10	16606.01	4886.09	76.98	77.27	23.02
5	493866.84	5373.03	1247.08	4125.95	26865.13	20731.96	6133.17	76.79	77.17	23.21
6	492609.37	5373.03	1257.47	4115.56	32238.15	24847.52	7390.63	76.60	77.07	23.40
7	491341.42	5373.03	1267.95	4105.08	37611.18	28952.60	8658.58	76.40	76.98	23.60
8	490062.91	5373.03	1278.51	4094.51	42984.20	33047.11	9937.10	76.21	76.88	23.80
9	488773.74	5373.03	1289.17	4083.86	48357.23	37130.97	11226.26	76.01	76.78	23.99
10	487473.83	5373.03	1299.91	4073.11	53730.26	41204.08	12526.17	75.81	76.69	24.19
11	486163.08	5373.03	1310.74	4062.28	59103.28	45266.36	13836.92	75.61	76.59	24.39
12	484841.42	5373.03	1321.67	4051.36	64476.31	49317.72	15158.59	75.40	76.49	24.60
13	483508.74	5373.03	1332.68	4040.35	69849.33	53358.07	16491.27	75.20	76.39	24.80
14	482164.95	5373.03	1343.79	4029.24	75222.36	57387.31	17835.05	74.99	76.29	25.01
15	480809.96	5373.03	1354.98	4018.04	80595.38	61405.35	19190.04	74.78	76.19	25.22
170	51347.53	5373.03	4904.26	468.77	913414.35	464761.89	448652.47	8.72	50.88	91.28
171	46402.41	5373.03	4945.13	427.90	918787.38	465189.78	453597.60	7.96	50.63	92.04
172	41416.07	5373.03	4986.34	386.69	824160.40	465576.47	458583.93	7.20	50.38	92.80
173	36388.17	5373.03	5027.89	345.13	929533.43	465921.60	463611.83	6.42	50.12	93.58
174	31318.38	5373.03	5069.79	303.23	934906.45	466224.84	468681.62	5.64	49.87	94.36
175	26206.34	5373.03	5112.04	280.99	940279.48	466485.82	473793.66	4.86	49.61	95.14
176	21051.71	5373.03	5154.64	218.39	945652.51	466704.21	478948.30	4.06	49.35	95.94
177	15854.11	5373.03	5197.59	175.43	951025.53	466879.64	484145.89	3.27	49.09	96.74
178	10613.20	5373.03	5240.91	132.12	956398.56	467011.76	489386.80	2.46	48.83	97.54
179	5328.62	5373.03	5284.58	88.44	961771.58	467100.20	494671.38	1.65	48.57	98.35
180	0.00	5373.03	5328.62	44.41	967144.61	467144.61	500000.00	0.83	48.30	99.17

t:	Month	E(t):	Equity Portion of the Payment	S(t):	Total Equity to Date
U(t):	Unpaid Balance	I(t):	Interest Charge Contained in the Payment	f(t):	Percentage of Interest Charge
A(t):	Monthly Payment	C(t):	Total Payment to Date	F(t):	Percentage of Cumulative Interest Charge

Example 9.1

Suppose a manufacturing productivity improvement project is to be financed by borrowing $500,000 from an industrial development bank. The annual nominal interest rate for the loan is 10%. The loan is to be repaid in equal monthly installments over a period of 15 years. The first payment on the loan is to be made exactly one month after financing is approved. A detailed analysis of the loan schedule is desired. Table 9.1 presents a partial listing of the loan repayment schedule. These figures are computed using the ENGINEA software, which is described in Chapter 12.

The tabulated result shows a monthly payment of $5,373.03 on the loan. If time $t = 10$ months, one can see the following results:

$U(10) = \$487,473.83$ (unpaid balance)
$A(10) = \$5,373.03$ (monthly payment)
$E(10) = \$1,299.91$ (equity portion of the tenth payment)
$I(10) = \$4,073.11$ (interest charge contained in the tenth payment)
$C(10) = \$53,730.26$ (total payment to date)

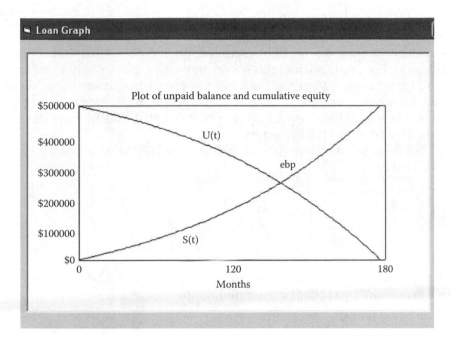

Figure 9.1 Plot of unpaid balance and cumulative equity.

$S(10)$ = \$12,526.17 (total equity to date)
$f(10)$ = 75.81% (percentage of the tenth payment going into interest charge)
$\Gamma(10)$ = 76.69% (percentage of the total payment going into interest charge)

Thus, over 76% of the sum of the first 10 installment payments goes into interest charges. The analysis shows that by time $t = 180$, the unpaid balance has been reduced to zero. That is, $U(180) = 0.0$. The total payment made on the loan is \$967,144.61 and the total interest charge is \$967,144.61 – \$500,000 = \$467,144.61. So, 48.30% of the total payment goes into interest charges. The information about interest charges might be very useful for tax purposes. The tabulated output shows that equity builds up slowly, whereas the unpaid balance decreases slowly. Note that very little equity is accumulated during the first 3 years of the loan schedule. This is shown graphically in Figure 9.1 (also constructed using the ENGINEA software). The effects of inflation, depreciation, property appreciation, and other economic factors are not included in the analysis just presented, but a project analysis should include such factors whenever they are relevant to the loan situation.

9.1.1 Equity break-even point

The point at which the curves intersect is referred to as the *equity break-even point*. It indicates when the unpaid balance is exactly equal to the accumulated equity or the cumulative principal payment. For the preceding example, the equity break-even point is approximately 120 months (10 years). The importance of the equity break-even point is that any equity accumulated after that point represents the amount of ownership or equity that the debtor is entitled to after the unpaid balance on the loan is settled with project collateral. The implication of this is very important, particularly in the case of mortgage loans. "Mortgage" is a word of French origin, meaning "death pledge," which, perhaps, is an ironic reference to the burden of mortgage loans. The equity break-even point can be calculated directly from the formula derived in the following text:

Let the equity break-even point, x, be defined as the point where $U(x) = S(x)$. That is,

$$A\left[\frac{1-(1+i)^{-(n-x)}}{i}\right] = A\left[\frac{(1+i)^x - 1}{i(1+i)^n}\right]$$

Multiplying both the numerator and denominator of the left-hand side of this expression by $(1+i)n$ and then simplifying yields

$$\frac{(1+i)^n - (1+i)^x}{i(1+i)^n}$$

on the left-hand side. Consequently, we have

$$(1+i)^n - (1+i)^x = (1+i)^x - 1$$

$$(1+i)^x = \frac{(1+i)^n + 1}{2}$$

which yields the equity break-even expression

$$x = \frac{\ln\left[0.5(1+i)^n + 0.5\right]}{\ln(1+i)}$$

where

$\quad ln$ = the natural log function
$\quad n$ = the number of periods in the life of the loan
$\quad i$ = the interest rate per period

Figure 9.2 presents a plot of the total loan payment and the cumulative equity with respect to time. The total payment starts from $0.0 at time 0 and

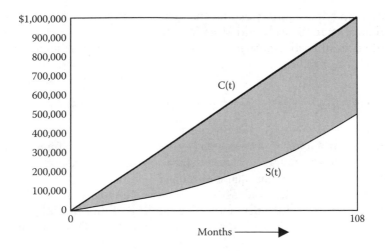

Figure 9.2 Plot of total loan payment and total equity.

goes up to $967,144.61 by the end of the last month of the installment pay-
ments. Because only $500,000 was borrowed, the total interest payment on
the loan is $967,144.61 − $500,000 = $467,144.61. The cumulative principal
payment starts at $0.0 at time 0 and slowly builds up to $500,000, which is
the original loan amount.

Figure 9.3 presents a plot of the percentage of interest charge in the
monthly payments and the percentage of interest charge in the total payment.
The percentage of interest charge in the monthly payments starts at 77.55%
for the first month and decreases to 0.83% for the last month. By comparison,
the percentage of interest in the total payment starts also at 77.55% for the
first month and slowly decreases to 48.30% by the time the last payment is
made at time 180. Table 9.1 and Figure 9.3 show that an increasing proportion

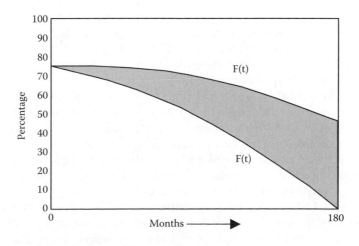

Figure 9.3 Plot of percentage of interest charge.

of the monthly payment goes into the principal payment as time goes on. If the interest charges are tax deductible, the decreasing values of $f(t)$ mean that there would be decreasing tax benefits from the interest charges in the later months of the loan.

9.2   Introduction to tent cash-flow analysis

Analysis of arithmetic gradient series (AGS) cash flows is one of the more convoluted problems in engineering economic analysis. This chapter presents an interesting graphical representation of AGS cash flows and the approach is in the familiar shape of tents.

Several designs as well as a closed-form analysis of AGS cash-flow profiles are presented for a better understanding of this and other cash-flow profiles and their analysis. A general tent equation (GTE) that can be used to solve various AGS cash flows is developed. The general equation can be used for various tent structures with the appropriate manipulations. GTE eliminates the problem of using the wrong number of periods and is amenable to software implementation.

9.3   Special application of AGS

AGS cash flows feature prominently in many contract payments, but misinterpretation of them can seriously distort the financial reality of a situation. Good examples can be found in the contracts of sports professionals. The pervasiveness of, and extensive publicity attending, such contracts make the analysis of arithmetic gradient series both appealing and economically necessary. A good example is the 1984 contract of Steve Young, a quarterback for the LA Express team in the former USFL (United States Football League). The contract was widely reported as being worth $40 million at that time. The cash-flow profile of the contract revealed an intricate use of various segments of arithmetic gradient series cash flows. When everything was taken into account, the $40 million touted in the press amounted only to a present worth of the contract of about $5 million at that time. The trick was that the club included some deferred payments stretching over 37 years (1990–2027) at a 1984 present cost of only $2.9 million. The deferred payments were reported as being worth $34 million, which was the raw sum of the amounts in the deferred cash-flow profile. Thus, it turns out that clever manipulation of an AGS cash flow can create unfounded perceptions of the worth of a professional sports contract. This may explain why some sports professionals end up almost bankrupt even after receiving what they assume to be multimillion-dollar contracts. Similar examples have been found in reviewing the contracts of other sports professionals.

Other real-life examples of gradient cash-flow constructions can be found in general investment cash flows and maintenance operations. In a particular economic analysis of maintenance operations, an escalation factor was applied to the annual cost of maintaining a major piece of equipment.

The escalation is justified because of inflation, increased labor costs, and other forecasted needs.

9.4   Design and analysis of tent cash-flow profiles

AGS cash flows usually start with some base amount at the end of the first period and then increase or decrease by a constant amount thereafter. The nonzero base amount is denoted as A_T starting at period T. The analysis of the present worth for such cash flows requires breaking the cash flow into a uniform series cash flow of amount A_T starting at period T and an AGS cash flow with a zero base amount. The uniform series present worth formula is used to calculate the present worth of the uniform series portion, whereas the basic AGS formula is used to calculate the arithmetic gradient series part of the cash-flow profile. The overall present worth is then calculated as follows:

$$P = P_{uniform\ series} \pm P_{arithmetic\ gradient\ series}$$

Figure 9.4 presents a conventional AGS cash flow. Each cash-flow amount at time t is defined as $A_t = (t{-}1)$. The standard formula for this basic AGS profile is derived as follows:

$$P = \sum_{t=1}^{n} A_t (1+i)^{-t}$$

$$= \sum_{t=1}^{n} (t-1)\, G\, (1+i)^{-t}$$

$$= G \sum_{t=1}^{n} (t-1)\, (1+i)^{-t}$$

$$= \dots\dots\dots\dots\dots$$

$$= G \left[\frac{(1+i)^{n} - (1+ni)}{i^{2}(1+i)^{n}} \right]$$

$$= G(P/G, i, n), \text{ in tabulated form}$$

The computational process of deriving and using the P/G formula is where new engineering economy students often stumble. Recognizing the tent-like structure of the cash flow, and the fact that several applications of AGS are beyond its conventional format, leads to the idea of tent cash flows. Figure 9.5 presents the basic tent (BT) cash-flow profile. It is composed of

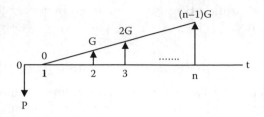

Figure 9.4 Conventional Arithmetic Gradient Series (AGS) cash flow.

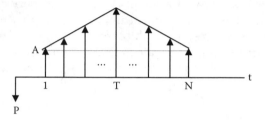

Figure 9.5 Basic Tent (BT) cash-flow profile.

an up-slope gradient and a down-slope portion, both on a uniform series base of $A = A_1$.

   Computational formula analysis of BT cash flow is shown here.

For the first half of the cash flow (from $t = 1$ to $t = T$),

$$P_a = \sum_{t=1}^{T} A_1(1+i)^{-t} + \sum_{t=1}^{T} A_t(1+i)^{-t}$$

where $A_t = (t-1)G$.

$$P_a = \sum_{t=1}^{T} A_1(1+i)^{-t} + G\sum_{t=1}^{T} (t-1)(1+i)^{-t}$$

$$P_a = A_1\left[\frac{(1+i)^T - 1}{i(1+i)^T}\right] + G\left[\frac{(1+i)^T - (1+Ti)}{i^2(1+i)^T}\right] \qquad (9.1)$$

For the second half of the cash flow (from $t = T+1$ to $t = N$),

$$P_b = \sum_{t=T+1}^{N} A_{T+1}(1+i)^{-t} - \sum_{t=T+1}^{N} A_t(1+i)^{-t}$$

where $A_t = (t-1)G$.

$$P_b = A_{T+1} \sum_{t=T+1}^{N} (1+i)^{-t} - \sum_{t=T+1}^{N} (t-1)G(1+i)^{-t}$$

$$\vdots$$

$$\vdots$$

$$P_b = A_{T+1} \left[\frac{(1+i)^{N-T}-1}{i(1+i)^{N-T}} \right] - G\left[\frac{(1+i)^{N-T}-[1+(N-T)i]}{i^2(1+i)^{N-T}} \right]$$

Let $x = (N - T)$,

$$\therefore P_b = A_{T+1} \left[\frac{(1+i)^x - 1}{i(1+i)^x} \right] - G\left[\frac{(1+i)^x - (1+xi)}{i^2(1+i)^x} \right] \qquad (9.2)$$

Hence,

$$P = P_a + P_b(1+i)^{-T} \qquad (9.3)$$

The basic approach to solving this type of cash-flow profile is to partition it into simpler forms. By partitioning, BT can be solved directly using standard cash-flow conversion factors. That is, the solution can be obtained as the sum of a uniform series cash flow (base amount), an increasing AGS, and a decreasing AGS. That is,

$$P = G(P/A,i,N) + G(P/G,i,T) + \left[G(T-2)(P/A,i,N-T) - G(P/G,i,N-T) \right](P/F,i,T)$$

The preceding two approaches should yield the same result. Obviously, the partitioning approach using existing standard factors is a more ingenious method. One common student error when using the existing AGS factor is the use of an incorrect number of periods, n. It should be recognized that the standard AGS factor was derived for a situation where P is located one period before the "nose" of the increasing series. Students often tend to locate P right at the same point on the time line as the "nose" of the series, which means that n will be off by one unit. One way to avoid this error is to redraw the time line and renumber it from a reference point of zero; that is, relocate time zero ($t = 0$) to one period before the AGS begins.

Figure 9.6 shows a profile of the ET cash-flow profile. It has a constant amount of increasing and decreasing AGS. The magnitudes of the cash-flow amounts at times T_j ($j = 1, 2, \ldots$) are equal.

For Part A:

$$P_A = P_1 + P_{T_1+1}(1+i)^{-T_1} \qquad (9.4)$$

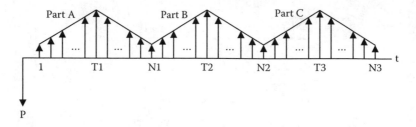

Figure 9.6 Executive tent (ET) cash-flow profile.

where

$$P_1 = A_1\left[\frac{(1+i)^{T_1}-1}{i(1+i)^{T_1}}\right]+G\left[\frac{(1+i)^{T_1}-(1+T_1 i)}{i^2(1+i)^{T_1}}\right] \tag{9.5}$$

and

$$P_{T_1+1} = A_{T_1+1}\left[\frac{(1+i)^{x_1}-1}{i(1+i)^{x_1}}\right]-G\left[\frac{(1+i)^{x_1}-(1+x_1 i)}{i^2(1+i)^{x_1}}\right] \tag{9.6}$$

while $x_1 = (N_1 - T_1)$.

For Part B:

$$P_B = P_{N_1+1}(1+i)^{-N_1} + P_{T_2+1}(1+i)^{-T_2} \tag{9.7}$$

where

$$P_{N_1+1} = A_{N_1+1}\left[\frac{(1+i)^{T_2}-1}{i(1+i)^{T_2}}\right]+G\left[\frac{(1+i)^{T_2}-(1+T_2 i)}{i^2(1+i)^{T_2}}\right] \tag{9.8}$$

and

$$P_{T_2+1} = A_{T_2+1}\left[\frac{(1+i)^{x_2}-1}{i(1+i)^{x_2}}\right]-G\left[\frac{(1+i)^{x_2}-(1+x_2 i)}{i^2(1+i)^{x_2}}\right] \tag{9.9}$$

while $x_2 = (N_2 - T_2)$.

For Part C:

$$P_C = P_{N_2+1}(1+i)^{-N_2} + P_{T_3+1}(1+i)^{-T_3} \tag{9.10}$$

where

$$P_{N_2+1} = A_{N_2+1}\left[\frac{(1+i)^{T_3}-1}{i(1+i)^{T_3}}\right] + G\left[\frac{(1+i)^{T_3}-(1+T_3 i)}{i^2(1+i)^{T_3}}\right] \qquad (9.11)$$

and

$$P_{T_3+1} = A_{T_3+1}\left[\frac{(1+i)^{x_3}-1}{i(1+i)^{x_3}}\right] - G\left[\frac{(1+i)^{x_3}-(1+x_3 i)}{i^2(1+i)^{x_3}}\right] \qquad (9.12)$$

while $x_3 = (N_3 - T_3)$.

$$\therefore P = P_A + P_B + P_C \qquad (9.13)$$

Figure 9.7 presents a Saw-Tooth Tent (STT) cash-flow profile. The present value analysis of the cash flow is computed as follows:

$$P = P_1 + P_{T_1+1}(1+i)^{-T_1} + P_{T_2+1}(1+i)^{-T_2} \qquad (9.14)$$

where

$$P_1 = A_1\left[\frac{(1+i)^{T_1}-1}{i(1+i)^{T_1}}\right] + G\left[\frac{(1+i)^{T_1}-(1+T_1 i)}{i^2(1+i)^{T_1}}\right] \qquad (9.15)$$

$$P_{T_1+1} = A_{T_1+1}\left[\frac{(1+i)^{T_2}-1}{i(1+i)^{T_2}}\right] + G\left[\frac{(1+i)^{T_2}-(1+T_2 i)}{i^2(1+i)^{T_2}}\right] \qquad (9.16)$$

$$P_{T_2+1} = A_{T_2+1}\left[\frac{(1+i)^{N}-1}{i(1+i)^{N}}\right] + G\left[\frac{(1+i)^{N}-(1+N i)}{i^2(1+i)^{N}}\right] \qquad (9.17)$$

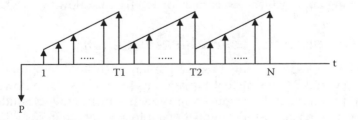

Figure 9.7 Saw-Tooth Tent (STT) cash-flow profile.

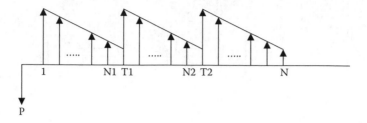

Figure 9.8 Reversed Saw-Tooth Tent (R-STT) cash-flow profile.

In Figure 9.8, a Reversed Saw-Tooth Tent (R-STT) is constructed. Its present value computation is handled as shown here.

$$P = P_1 + P_{T_1}(1+i)^{-T_1} + P_{T_2}(1+i)^{-T_2} \tag{9.18}$$

where

$$P_1 = A_1\left[\frac{(1+i)^{N_1} - 1}{i(1+i)^{N_1}}\right] - G\left[\frac{(1+i)^{N_1} - (1+N_1 i)}{i^2(1+i)^{N_1}}\right] \tag{9.19}$$

$$P_{T_1} = A_{T_1}\left[\frac{(1+i)^{x_1} - 1}{i(1+i)^{x_1}}\right] - G\left[\frac{(1+i)^{x_1} - (1+x_1 i)}{i^2(1+i)^{x_1}}\right] \tag{9.20}$$

while $x_1 = (N_2 - N_1)$.

$$P_{T_2} = A_{T_2}\left[\frac{(1+i)^{x_2} - 1}{i(1+i)^{x_2}}\right] - G\left[\frac{(1+i)^{x_2} - (1+x_2 i)}{i^2(1+i)^{x_2}}\right] \tag{9.21}$$

while $x_2 = (N - N_2)$.

Increasingly complicated profiles can be designed depending on the level of complexity desired to test different levels of student understanding. Note that the ET, STT, and R-STT cash flows can also be solved by the partitioning approach shown earlier for the BT cash flow.

9.5   Derivation of general tent equation

To facilitate a less convoluted use of the tent cash-flow computations, this section presents a General Tent Equation (GTE) for AGS cash flows. This is suitable for adoption by engineering economy instructors or students. We combine all the tent cash-flow equations into a general tent cash-flow equation that is amenable to software implementation. The general equation is as follows:

$$P_0 = A_0 + A_1 \left[\frac{(1+i)^T - 1}{i(1+i)^T} \right] + G_1 \left[\frac{(1+i)^T - (1+Ti)}{i^2(1+i)^T} \right] \qquad (9.22)$$

$$P_T = A_{T_1} \left[\frac{(1+i)^x - 1}{i(1+i)^x} \right] + G_2 \left[\frac{(1+i)^x - (1+xi)}{i^2(1+i)^x} \right] \qquad (9.23)$$

when $G_1 \geq 0$ or $G_2 \leq 0$,

$$TPV_0 = P_0 + P_T (1+i)^{-T} \qquad (9.24)$$

when $G_1 < 0$ or $G_2 > 0$,

$$TPV_0 = P_0 - P_T (1+i)^{-T} \qquad (9.25)$$

where

TPV_0 = Total Present Value at time t = 0
P_0 = Present Value for the first half of the tent at time t = 0
P_T = Present Value for the second half of the tent at time t = T
i = interest rate in fractions
N = number of periods
T = the center time value of the tent
$x = (N - T)$
A_0 = amount at time t = 0
A_1 = amount at time t = 1
A_{T1} = amount at time = T + 1
G_1 = gradient series of the first half of the tent
   $\Rightarrow$ increasing G_1 (up-slope) is a positive value
      and decreasing G_1 (down-slope) is a negative value
G_2 = gradient series of the first half of the tent
   $\Rightarrow$ increasing G_2 (up-slope) is a positive value
      and decreasing G_2 (down-slope) is a negative value

   This GTE is based on the BT but can be used for either one-sided (see Figure 9.1) or two-sided (see Figure 9.2) tents. It can also be used for Executive Tents (ET) with more than two cycles. The process involved in this case is to divide the tent into smaller sections of two cycles, such as Part A, Part B, and Part C (see Figure 9.3), and then to use the equation to solve for the PV of each part at time $t = 0$, $t = N\,1$, and $t = N\,2$. The PV of the future values at $t = N\,1$ and $t = N\,2$ are added to the PV at $t = 0$ to determine the overall PV of the ET at $t = 0$. This GTE eliminates the problem associated with having an incorrect number of periods and makes AGS analysis interesting. This equation can also be used for situations where the uniform series base of A_1 is zero by finding the PV at time $t = 1$ and taking it back one step to determine the total present value at time $t = 0$. Other scenarios can also be considered by appropriate manipulations of the equation.

Example 9.2

Using the GTE, find the PV of the multiple BTs (Figure 9.9) at time t = 0, where $i = 10\%$ per period.

    Solving this tent cash flow usually poses a lot of problems for students; however, the use of the GTE reduces such problems considerably. Because this tent consists of two BTs, we will use the general equation to determine the present value at time t = 0 for the first BT and the PV at time $t = 11$, we will then find the PV at time $t = 0$ for the second tent and sum it to the PV for the first tent and the PV at time $t = 1$.

    Using the GTE for the first basic tent, we obtained $P_0 = \$101.17$ for the first half of the first basic tent, and $P_5 = \$82.24$ for the second half of the tent: $PV_0 = \$152.23$.

    To solve the second BT, we renumbered the tent so that $t = 11$ becomes $t = 0$, $t = 12$ becomes $t = 1$, and so on. Therefore, $t = T$ is at $t = 7$, and we obtained $P_{11} = \$88.16$ for the first half of the second BT and $P_{18} = \$82.24$ for the second half of the second BT: $PV_{11} = \$130.36$.

    The total PV for the dual BT at time $t = 0$ is

$$TNPV_0 = P_{-1}(F/P, i\%, N) + P_0 + P_{11}\left(P/F, i\%, N\right)$$

$$TNPV_0 = 5(F/P, 10\%, 1) + 152.23 + 130.36\left(P/F, 10\%, 11\right)$$

$$TNPV_0 = \$203.42$$

Therefore, the overall PV of this dual BT at time $t = 0$ is **\$203.42.**

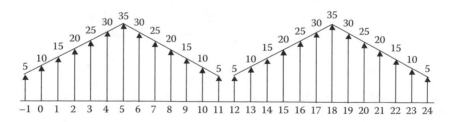

Figure 9.9 Dual BT cash-flows computation.

chapter ten

Multiattribute investment analysis and selection

This chapter presents useful techniques for assessing and comparing investments in order to improve the selection process. The techniques presented include utility models, the project value model, polar plots, benchmarking techniques, and the analytic hierarchy process (AHP).

10.1 The problem of investment selection

Investment selection is an important aspect of investment planning. The right investment must be undertaken at the right time to satisfy the constraints of time and resources. A combination of criteria can be used to help in investment selection, including technical merit, management desire, schedule efficiency, benefit/cost ratio, resource availability, criticality of need, availability of sponsors, and user acceptance.

Many aspects of investment selection cannot be expressed in quantitative terms. For this reason, investment analysis and selection must be addressed by techniques that permit the incorporation of both quantitative and qualitative factors. Some techniques for investment analysis and selection are presented in the sections that follow. These techniques facilitate the coupling of quantitative and qualitative considerations in the investment decision process. Such techniques as net present value, profit ratio, and equity break-even point, which have been presented in the preceding chapters, are also useful for investment selection strategies.

10.2 Utility models

The term *utility* refers to the rational behavior of a decision maker faced with making a choice in an uncertain situation. The overall utility of an investment can be measured in terms of both quantitative and qualitative factors. This section presents an approach to investment assessment based on utility models that have been developed within an extensive body of literature. The approach fits an empirical utility function to each factor that is to be included in a multiattribute selection model. The specific utility values (weights) that are obtained from the utility functions are used as the basis for selecting an investment.

Utility theory is a branch of decision analysis that involves the building of mathematical models to describe the behavior of a decision maker faced with making a choice among alternatives in the presence of risk. Several utility models are available in the management science literature. The utility of a composite set of outcomes of n decision factors is expressed in the following general form:

$$U(x) = U(x_1, x_2, \ldots, x_n)$$

where x_i = specific outcome of attribute X_i, i = 1, 2, ..., n and $U(x)$ is the utility of the set of outcomes to the decision maker. The basic assumption of utility theory is that people make decisions with the objective of maximizing those decisions' *expected utility*. Drawing on an example presented by Park and Sharp-Bette (1990), we may consider a decision maker whose utility function with respect to investment selection is represented by the following expression:

$$u(x) = 1 - e^{-0.0001x}$$

where x represents a measure of the benefit derived from an investment. Benefit, in this sense, may be a combination of several factors (e.g., quality improvement, cost reduction, or productivity improvement) that can be represented in dollar terms. Suppose this decision maker is faced with a choice between two investment alternatives, each of which has benefits specified as follows:

Investment I: Probabilistic levels of investment benefits

Benefit, x	$10,000	$0	$10,000	$20,000	$30,000
Probability, $P(x)$	0.2	0.2	0.2	0.2	0.2

Investment II: A definite benefit of $5,000
Assuming an initial benefit of zero and identical levels of required investment, the decision maker must choose between the two investments. For Investment I, the expected utility is computed as follows:

$$E[u(x)] = \sum u(x)\{P(x)\}$$

Benefit, x	Utility, $u(x)$	$P(x)$	$u(x)\,P(x)$
$10,000	1.7183	0.2	0.3437
$0	0	0.2	0
$10,000	0.6321	0.2	0.1264
$20,000	0.8647	0.2	0.1729
$30,000	0.9502	0.2	0.1900
		Sum	0.1456

Thus, $E[u(x)_1] = 0.1456$. For Investment II, we have $u(x)_2 = u(\$5,000) = 0.3935$. Consequently, the investment providing the certain amount of $5000 is preferred to the riskier Investment I, even though Investment I has a higher expected benefit of $\Sigma x P(x) = \$10,000$. A plot of the utility function used in the preceding example is presented in Figure 10.1.

If the expected utility of 0.1456 is set equal to the decision-maker's utility function, we obtain the following:

$$0.1456 = 1 - e^{-0.0001x^*}$$

which yields $x^* = \$1,574$, referred to as the *certainty equivalent* (CE) of Investment I (CE$_1$ = 1,574). The certainty equivalent of an alternative with variable outcomes is a *certain amount* (CA), which a decision maker will consider to be desirable to the same degree as the variable outcomes of the alternative. In general, if CA represents the certain amount of benefit that can be obtained from Investment II, then the criteria for making a choice between the two investments can be summarized as follows:

If CA < $1,574, select Investment I.
If CA = $1,574, select either investment.
If CA > $1,574, select Investment II.

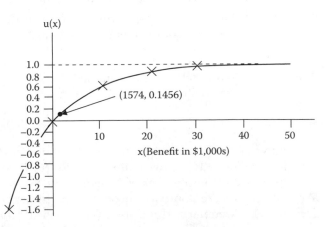

Figure 10.1 Utility function and certainty equivalent.

The key in using utility theory for investment selection is choosing the proper utility model. The sections that follow describe two simple but widely used utility models: the *additive utility model* and the *multiplicative utility model*.

10.2.1 Additive utility model

The additive utility of a combination of outcomes of n factors $(X_1, X_2, ..., X_n)$ is expressed as follows:

$$U(x) = \sum_{i=1}^{n} U\left(x_i, \bar{x}_i^0\right)$$

$$= \sum_{i=1}^{n} k_i U(x_i)$$

where

x_i = measured or observed outcome of attribute i

n = number of factors to be compared

x = combination of the outcomes of n factors

$U(x_i)$ = utility of the outcome for attribute i, x_i

$U(x)$ = combined utility of the set of outcomes, x

k_i = weight or scaling factor for attribute $i (0 < k_i < 1)$

X_i = variable notation for attribute i

x_i^0 = worst outcome of attribute i

x_i^* = best outcome of attribute i

$\bar{x}_i^0$ = set of worst outcomes for the complement of x_i

$U(x_i, \bar{x}_i^0)$ = utility of the outcome of attribute I and the set of worst outcomes for the complement of attribute i

$k_i = U(x_i^*, \bar{x}_i^0)$

$\sum_{i=1}^{n} k_i$ = 1.0 (required for the additive model).

Example 10.1

Let **A** be a collection of four investment attributes defined as **A** = {Profit, Flexibility, Quality, Productivity}. Now define **X** = {Profit, Flexibility} as a subset of A. Then, X is the complement of **X** defined as X = {Quality, Productivity}. An example of the comparison of two investments under the additive utility model is summarized in Table 10.1 and yields the following results:

Table 10.1 Example of Additive Utility Model

Attribute (i)	Weight (k_i)	Investment A $U_i(x_i)$	Investment B $U_i(x_i)$
Profitability	0.4	0.95	0.90
Flexibility	0.2	0.45	0.98
Quality	0.3	0.35	0.20
Throughput	0.1	0.75	0.10
	1.00		

$$U(x)_A = \sum_{i=1}^{n} k_i U_i(x_i) = .4(.95) + .2(.45) + .3(.35) + .1(.75) = 0.650$$

$$U(x)_B = \sum_{i=1}^{n} k_i U_i(x_i) = .4(.90) + .2(.98) + .3(.20) + .1(.10) = 0.626$$

Because $U(x)_A > U(x)_B$, Investment A is selected.

10.2.2   Multiplicative utility model

Under the multiplicative utility model, the utility of a combination of outcomes of n factors $(X_1, X_2, \ldots, X_n)$ is expressed as

$$U(x) = \frac{1}{C}\left[\prod_{i=1}^{n} \left(C k_i U_i(x_i) + 1 \right) - 1 \right]$$

where C and k_i are scaling constants satisfying the following conditions:

$$\prod_{i=1}^{n} \left(1 + C k_i \right) - C = 1.0$$

$$1.0 < C < 0.0$$

$$0 < k_i < 1$$

The other variables are as defined previously for the additive model. Using the multiplicative model for the data in Table 10.1 yields $U(x)_A = 0.682$ and $U(x)_B = 0.676$. Thus, Investment A is the best option.

10.2.3   Fitting a utility function

An approach presented in this section for multiattribute investment selection is to fit an empirical utility function to each factor to be considered in the selection process. The specific utility values (weights) that are obtained from the utility functions may then be used in any of the standard investment justification methodologies. One way to develop empirical utility function for an investment attribute is to plot the "best" and "worst" outcomes expected from the attribute and then to fit a reasonable approximation of the utility function using concave, convex, linear, S-shaped, or any other logical functional form.

Alternately, if an appropriate probability density function can be assumed for the outcomes of the attribute, then the associated cumulative distribution function may yield a reasonable approximation of the utility values between 0 and 1 for corresponding outcomes of the attribute. In that case, the cumulative distribution function gives an estimate of the cumulative utility associated with increasing levels of attribute outcome. Simulation experiments, histogram plotting, and goodness-of-fit tests may be used to determine the most appropriate density function for the outcomes of a given attribute. For example, the following five attributes are used to illustrate how utility values may be developed for a set of investment attributes. The attributes are return on investment (ROI), productivity improvement, quality improvement, idle-time reduction, and safety improvement.

Example 10.2

Suppose we have historical data on the ROI for investing in a particular investment. Assume that the recorded ROI values range from 0 to 40%. Thus, the worst outcome is 0%, and the best outcome is 40%. A frequency distribution of the observed ROI values is developed and an appropriate probability density function (pdf) is fitted to the data. For our example, suppose the ROI is found to be exceptionally distributed with a mean of 12.1%. That is,

$$f(x) = \begin{cases} \dfrac{1}{\beta} e^{-x/\beta}, & \text{if } x \geq 0 \\ 0, & \text{otherwise} \end{cases}$$

$$F(x) = \begin{cases} 1 - e^{-x/\beta}, & \text{if } x \geq 0 \\ 0, & \text{otherwise} \end{cases}$$

$$\approx U(x)$$

where $\beta = 12.1$, $F(x)$ approximates $U(x)$. The probability density function and cumulative distribution function are shown graphically in Figure 10.2. The utility of any observed ROI within the applicable range may be read directly from the cumulative distribution function.

For the productivity improvement attribute, suppose it is found (based on historical data analysis) that the level of improvement is normally distributed with a mean of 10% and a standard deviation of 5%. That is,

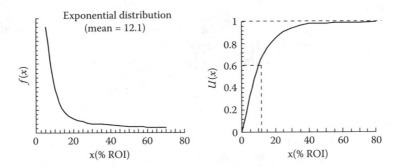

Figure 10.2 Estimated utility function for investment ROI.

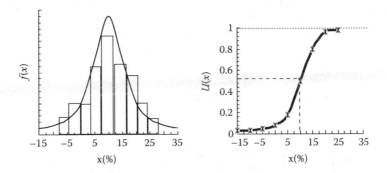

Figure 10.3 Utility function for productivity improvement.

$$f(x) = \frac{1}{\sqrt{2\pi}\sigma} e^{-\frac{1}{2}\left(\frac{x-\mu}{\sigma}\right)^2}, \quad -\infty < x < \infty$$

where $\pi = 10$ and $\sigma = 5$. Because the normal distribution does not have a closed-form expression for $F(x)$, $U(x)$ is estimated by plotting representative values based on the standard normal table. Figure 10.3 shows $f(x)$ and the estimated utility function for productivity improvement. The utility of productivity improvement may also be evaluated on the basis of cost reduction.

Suppose quality improvement is subjectively assumed to follow a beta distribution with shape parameters $\alpha = 1.0$ and $\beta = 2.9$. That is,

$$f(x) = \frac{\Gamma(\alpha+\beta)}{\Gamma(\alpha)\Gamma(\beta)} \cdot \frac{1}{(b-a)^{\alpha+\beta-1}} \cdot (x-a)^{\alpha-1}(b-x)^{\beta-1}$$

for $a \le x \le b$ and $\alpha > 0, \beta > 0$

where

a = lower limit for the distribution
b = upper limit for the distribution
α, β = the shape parameters for the distribution

As with the normal distribution, there is no closed-form expression for $F(x)$ for the beta distribution. However, if either of the shape parameters is a positive integer, then a binomial expansion can be used to obtain $F(x)$. Figure 10.4 shows a plot of $f(x)$ and the estimated $U(x)$ for quality improvement due to the proposed investment.

Based on work analysis observations, suppose idle-time reduction is found to be best described by a log normal distribution with a mean of 10% and standard deviation of 5%. This is represented as follows:

$$f(x) = \frac{1}{\sqrt{2\pi}\sigma} e^{-\frac{1}{2}\left(\frac{x-\mu}{\sigma}\right)^2}, \quad -\infty < x < \infty$$

There is no closed-form expression for $F(x)$. Figure 10.5 shows $f(x)$ and the estimated $U(x)$ for idle-time reduction due to the investment.

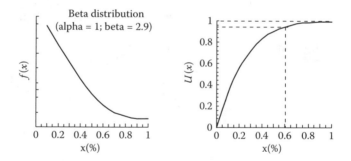

Figure 10.4 Utility function for quality improvement.

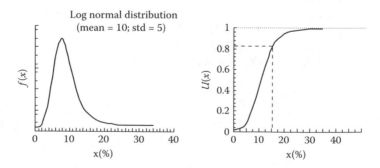

Figure 10.5 Utility function for idle-time reduction.

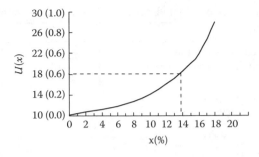

Figure 10.6 Utility function for safety improvement.

For the example, suppose safety improvement is assumed to have a previously known utility function, defined as follows:

$$U_p(x) = 30 - \sqrt{400 - x^2}$$

where x represents percent improvement in safety. For the expression, the unscaled utility values range from 10 (for 0% improvement) to 30 (for 20% improvement). To express any particular outcome of an attribute i, x_i, on a scale of 0.0 to 1.0, it is expressed as a proportion of the range of best to worst outcomes as follows:

$$X = \frac{x_i - x_i^0}{x_i^* - x_i^0}$$

where

X = outcome expressed on a scale of 0.0 to 1.0
x_i = measured or observed raw outcome of attribute i
x_i^0 = worst raw outcome of attribute i
x_i^* = best raw outcome of attribute i

The utility of the outcome may then be represented as $U(X)$ and read off of the empirical utility curve. Using the preceding approach, the utility function for safety improvement is scaled from 0.0 to 1.0. This is shown in Figure 10.6. The numbers within parentheses represent the scaled values. The respective utility values for the five attributes may be viewed as relative weights for comparing investment alternatives. The utility obtained from the modeled functions can be used in the additive and multiplicative utility models discussed earlier. For example, Table 10.2 shows a composite utility profile for a proposed investment.

Using the additive utility model, the *composite utility* (CU) of the investment, based on the five attributes, is given by

Table 10.2 Composite Utility for a Proposed Investment

Attribute (i)	k_i	Value	$U_i(x_i)$
Return on investment	0.30	12.1%	0.61
Productivity improvement	0.20	10.0%	0.49
Quality improvement	0.25	60.0%	0.93
Idle-time reduction	0.15	15.0%	0.86
Safety improvement	0.10	15.0%	0.40
	1.00		

$$U(X) = \sum_{i=1}^{n} k_i U_i(x_i)$$

$$= .30(.61) + .20(.49) + .25(.93) + .15(.86) + .10(.40) = 0.6825$$

This composite utility value may then be compared with the utilities of other investments. On the other hand, a single investment may be evaluated independently on the basis of some minimum acceptable level of utility (MALU) desired by the decision maker. The criteria for evaluating an investment based on MALU may be expressed by the following rule:

> *Investment j is acceptable if its composite utility, U(X)$_j$, is greater than MALU.*
> *Investment j is not acceptable if its composite utility, U(X)$_j$, is less than MALU.*

The utility of an investment may be evaluated on the basis of its economic, operational, or strategic importance to an organization. Utility functions can be incorporated into existing justification methodologies. For example, in the analytic hierarchy process, utility functions can be used to generate values that are, in turn, used to evaluate the relative preference levels of attributes and alternatives. Utility functions can be used to derive component weights when the overall effectiveness of investments is being compared. Utility functions can generate descriptive levels of investment performance, as well as indicate the limits of investment effectiveness, as shown by the S-curve in Figure 10.7.

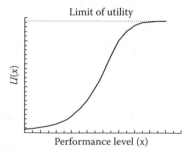

Figure 10.7 S-curve model for investment utility.

10.2.4   Investment value model

A technique that is related to utility modeling is the *investment value model* (IVM), which is an adaptation of the *manufacturing system value* (MSV) *model* presented by Troxler and Blank (1989). The model is suitable for the incorporation of utility values, and an example is presented here. The model provides a heuristic decision aid for comparing investment alternatives. "Value" is represented as a determined vector function that indicates the value of tangible and intangible "attributes" that characterize an alternative. Value can be expressed as follows:

$$V = f\left(A_1, A_2, \ldots, A_p\right)$$

where V = value, $A = (A_1, \ldots, A_n)$ = the vector of quantitative measures or attributes, and p = the number of attributes that characterize the investment. Examples of investment attributes include quality, throughput, capability, productivity, and cost performance. Attributes are considered to be a combined function of "factors," x_i, expressed as

$$A_k\left(x_1, x_2, \ldots, x_{m_k}\right) = \sum_{i=1}^{m_k} f_i\left(x_i\right)$$

where $\{x_i\}$ = the set of m factors associated with attribute A_k ($k = 1, 2, \ldots, p$), and f_i = the contribution function of factor x_i to attribute A_k. Examples of factors are market share, reliability, flexibility, user acceptance, capacity utilization, safety, and design functionality. Factors are themselves considered to be composed of "indicators," v_i, expressed as

$$x_i\left(v_1, v_2, \ldots, v_n\right) = \sum_{j=1}^{n} z_i\left(v_i\right)$$

where $\{v_j\}$ = the set of n indicators associated with factor x_i ($i = 1, 2, \ldots, m$) and z_j = the scaling function for each indicator variable v_j. Examples of indicators are debt ratio, investment responsiveness, lead time, learning curve, and scrap volume. By combining the above definitions, a composite measure of the value of an investment is given by the following:

$$PV = f\left(A_1, A_2, \ldots, A_p\right)$$

$$= f\left\{\left[\sum_{i=1}^{m_1} f_i\left(\sum_{j=1}^{n} z_j(v_j)\right)\right]_1, \left[\sum_{i=1}^{m_2} f_i\left(\sum_{j=1}^{n} z_j(v_j)\right)\right]_2, \ldots, \left[\sum_{i=1}^{m_k} f_i\left(\sum_{j=1}^{n} z_j(v_j)\right)\right]_p\right\}$$

Table 10.3 Comparison of Technology Values

Alternatives	Suitability ($k = 1$)	Capability ($k = 2$)	Performance ($k = 3$)	Productivity ($k = 4$)
Investment A	0.12	0.38	0.18	0.02
Investment B	0.30	0.40	0.28	1.00
Investment C	0.53	0.33	0.52	1.10

where *m* and *n* may assume different values for each attribute. A subjective measure to indicate the utility of the decision maker may be included in the model by using an attribute weighting factor, w_i, to obtain the following:

$$PV = f\left(w_1 A_1, w_2 A_2, \ldots, w_p A_p\right)$$

where

$$\sum_{k=1}^{p} w_k = 1, \quad \left(0 \le w_k \le 1\right)$$

As an example, an analysis using the preceding model to compare three investments on the basis of four attributes is presented in Table 10.3. The four attributes — *capability, suitability, performance,* and *productivity* — require careful interpretation before relative weights for the alternatives can be developed.

10.2.4.1   Capability
The term *capability* refers to the ability of equipment to produce certain features. For example, a certain piece of equipment may only produce horizontal or vertical slots, flat finishes, and so on. But a multiaxis machine can produce spiral grooves, internal metal removal from prismatic or rotational parts, thus increasing the part variety that can be made. In Table 10.3, the levels of increase in part variety from the three competing investments are 38, 40, and 33%, respectively.

10.2.4.2   Suitability
Suitability refers to the appropriateness of the investment to company operations. For example, chemical milling is more suitable for making holes in thin, flat metal sheets than drills. Drills need special fixtures to hold the thin metal down and protect it from wrinkling and buckling. The parts that the three investments are suitable for are, respectively, 12, 30, and 53% of the current part mix.

10.2.4.3   Performance
Performance, in this context, refers to the ability of the investment to produce high-quality outputs, or the ability to meet extra-tight performance

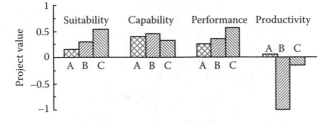

Figure 10.8 Relative system value weights of three alternatives.

requirements. In our example, the three investments can, respectively, meet tightened standards on 18, 28, and 52% of the normal set of jobs.

10.2.4.4 Productivity

Productivity can be measured by a simulation of the performance of the current system with the proposed technology at its current production rate, quality level, and part application. For the example in Table 10.3, the three investments, respectively, show increases of 0.02, 1.0, and 1.1 on a uniform scale of productivity measurement.

A plot of the histograms of the respective "values" of the three investments is shown in Figure 10.8. Investment C is the best alternative in terms of suitability and performance. Investment B shows the best capability measure, but its productivity is too low to justify the needed investment. Investment A, then, offers the best productivity, but its suitability measure is low.

The relative weights used in many justification methodologies are based on subjective propositions of the decision makers. Some of those subjective weights can be enhanced by the incorporation of utility models. For example, the weights shown in Table 10.3 could be obtained from utility functions.

10.2.5 Polar plots

Polar plots provide a means of visually comparing investment alternatives (Badiru, 1991). In a conventional polar plot, as shown in Figure 10.9, the vectors drawn from the center of the circle are on individual scales based on the outcome ranges for each attribute. For example, the vector for NPV (Net Present Value) is on a scale of $0 to $500,000, whereas the scale for Quality is from 0 to 10. It should be noted that the overall priority weights for the alternatives are not proportional to the areas of their respective polyhedrons.

A modification of the basic polar plot is presented in this section. The modification involves a procedure that normalizes the areas of the polyhedrons with respect to the total area of the base circle. With this modification, the normalized areas of the polyhedrons are proportional to the respective priority weights of the alternatives, so, the alternatives can be ranked on the basis of the areas of the polyhedrons. The steps involved in the modified approach are presented here:

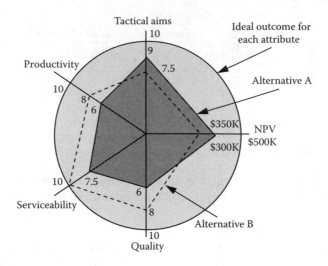

Figure 10.9 Basic polar plot.

1. Let n be the number of attributes involved in the comparison of alternatives, such that $n \geq 4$. Number the attributes in a preferred order $(1, 2, 3, \ldots, n)$.
2. If the attributes are considered to be equally important (i.e., equally weighted), compute the sector angle associated with each attribute as

$$\theta = \frac{360°}{n}$$

3. Draw a circle with a large enough radius. A radius of 2 in is usually adequate.
4. Convert the outcome range for each attribute to a standardized scale of 0 to 10 using appropriate transformation relationships.
5. For Attribute 1, draw a vertical vector up from the center of the circle to the edge of the circle.
6. Measure θ clockwise and draw a vector for Attribute 2. Repeat this step for all attributes in the numbered order.
7. For each alternative, mark its standardized relative outcome with respect to each attribute along the attribute's vector. If a 2-in radius is used for the base circle, then we have the following linear transformation relationship:

$$0.0 \text{ inches} = \text{rating score of } 0.0$$

$$2.0 \text{ inches} = \text{rating score of } 10.0$$

8. Connect the points marked for each alternative to form a polyhedron. Repeat this step for all alternatives.

9. Compute the area of the base circle as follows:

$$\Omega = \pi r^2$$

$$= 4\pi \quad \text{squared inches}$$

$$= 100\pi \quad \text{squared rating units}$$

10. Compute the area of the polyhedron corresponding to each alterna-
    tive. This can be done by partitioning each polyhedron into a set of
    triangles and then calculating the areas of the triangles. To calculate
    the area of each triangle, note that we know the lengths of two sides
    of the triangle and the angle subtended by the two sides. With these
    three known values, the area of each triangle can be calculated
    through basic trigonometric formulas.

    For example, the area of each polyhedron may be represented as
    λ_I ($I = 1, 2, ..., m$), where m is the number of alternatives. The area
    of each triangle in the polyhedron for a given alternative is then
    calculated as

$$\Delta_t = \frac{1}{2}(L_j)(L_{j+1})(Sin\theta)$$

where
   L_j = standardized rating with respect to attribute j
   L_{j+1} = standardized rating with respect to attribute j+1
   L_j and L_{j+1} are the two sides that subtend θ

Because $n \geq 4$, θ will be between 0 and 90 degrees, and $\sin(\theta)$ will be
strictly increasing over that interval.

   The area of the polyhedron for alternative i is then calculated as

$$\lambda_i = \sum_{t(i)=1}^{n} \Delta_{t(i)}$$

Note that θ is constant for a given number of attributes, and the area
of the polyhedron will be a function of the adjacent ratings (L_j and
L_{j+1}) only.

11. Compute the standardized area corresponding to each alternative as

$$w_i = \frac{\lambda_i}{\Omega}(100\%)$$

12. Rank the alternatives in decreasing order of λ_i. Select the highest
    ranked alternative as the preferred alternative.

Table 10.4 Ranges of Raw Evaluation Ratings for Polar Plots

Attribute (j)	Description	Rank k_j	Evaluation Range Lower Limit a_j	Upper Limit b_j
I	Quality	1	0.5	9
II	Profit (× $1,000)	2	0	100
III	Productivity	3	1	10
IV	Flexibility	4	0	12
V	Satisfaction	5	0	10

Table 10.5 Raw Evaluation Ratings for Modified Polar Plots

Alternatives	Attributes I ($j = 1$)	II ($j = 2$)	III ($j = 3$)	IV ($j = 4$)	V ($j = 5$)
A ($i = 1$)	5	50	3	6	10
B ($i = 2$)	1	20	1.5	9	2
C ($i = 3$)	8	75	4	11	1

Example 10.3

The problem presented here is used to illustrate how modified polar plots can be used to compare investment alternatives. Table 10.4 presents the ranges of possible evaluation ratings within which an alternative can be rated with respect to each of five attributes. The evaluation rating of an alternative with respect to attribute j must be between the given range a_j to b_j. Table 10.5 presents the data for raw evaluation ratings of three alternatives with respect to the five attributes specified in Table 10.4.

The attributes of Quality (I), Profit (II), and Productivity (III) are quantitative measures that can be objectively determined. The attributes of Flexibility (IV) and Customer Satisfaction (V) are subjective measures that can be intuitively rated by an experienced investment analyst. The steps in the solution are presented below.

Step 1: It is given that $n = 5$. The attributes are numbered in the following preferred order:
Quality: Attribute I
Profit: Attribute II
Productivity: Attribute III
Flexibility: Attribute IV
Satisfaction: Attribute V

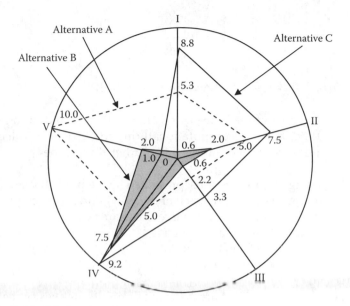

Figure 10.10 Modified polar plot.

Table 10.6 Standardized Evaluation Ratings for Modified Polar Plots

	Attributes				
	I	II	III	IV	V
Alternatives	($j = 1$)	($j = 2$)	($j = 3$)	($j = 4$)	($j = 5$)
A ($i = 1$)	5.3	5.0	2.2	5	10.0
B ($i = 2$)	0.6	2.0	0.6	7.5	2.0
C ($i = 3$)	8.8	7.5	3.3	9.2	1.0

Step 2:   The sector angle is computed as

$$\theta = 360°/n$$

$$= 72°$$

Step 3:   This step is shown in Figure 10.10.

Step 4:   Let Y_{ij} be the raw evaluation rating of alternative i with respect to attribute j (see Table 10.5).
Let Z_{ij} be the standardized evaluation rating.
The standardized evaluation ratings (between 0.0 and 10.0) shown in Table 10.6 were obtained by using the following linear transformation relationship:

$$Z_{ij} = 10 \left[\frac{(Y_{ij} - a_j)}{b_j - a_j} \right]$$

Steps 5, 6, 7, 8: These are shown in Figure 10.10.

Step 9:   The area of the base circle is $\Omega = 100\pi$ squared rating units. Note that it is computationally more efficient to calculate the areas in terms of rating units rather than inches.

Step 10: Using the expressions presented in Step 10, the areas of the triangles making up each of the polyhedrons are computed and summed up. The respective areas are
$\lambda_A = 72.04$ squared units
$\lambda_B = 10.98$ squared units
$\lambda_C = 66.14$ squared units

Step 11: The standardized areas for the three alternatives are as follows:
$w_A = 22.93\%$
$w_B = 3.50\%$
$w_C = 21.05\%$

Step 12: On the basis of the standardized areas in Step 11, Alternative A is found to be the best choice.

As an extension to the modification just presented, the sector angle may be a variable indicating relative attribute weights, whereas the radius represents the evaluation rating of the alternatives with respect to the weighted attribute. That is, if the attributes are not equally weighted, the sector angles will not all be equal. In that case, the sector angle for each attribute is computed as

$$\theta_j = p_j(360°)$$

where p_j = the relative numeric weight of each of n attributes

$$\sum_{j=1}^{n} p_j = 1.0$$

Suppose the attributes in the preceding example are considered to have unequal weights, as shown in Table 10.7.

Table 10.7 Relative Weighting of Attributes for Polar Plots

Attribute (i)	Weight p_j	Angle θ_j
I	0.333	119.88
II	0.267	96.12
III	0.200	72.00
IV	0.133	47.88
V	0.067	24.12
	1.000	360.00

The resulting polar plots for weighted sector angles are shown in Figure 10.11. The respective weighted areas for the alternatives are

λ_A = 51.56 squared units
λ_B = 9.07 squared units
λ_C = 60.56 squared units

The standardized areas for the alternatives are as follows:

w_A = 16.41%
w_B = 2.89%
w_C = 19.28%

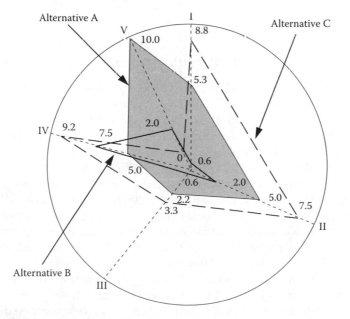

Figure 10.11 Polar plot with weighted sector angles.

Thus, if the given attributes are weighted as shown in Table 10.7, Alternative C will turn out to be the best choice. However, it should be noted that the relative weights of the attributes are too skewed, resulting in some sector angles being greater than 90 degrees. It is preferable to have the attribute weights assigned in such a way that all sector angles are less than 90 degrees. This leads to more consistent evaluation because $\sin(\theta)$ is strictly increasing between 0 and 90 degrees.

It should also be noted that the weighted areas for the alternatives are sensitive to the order in which the attributes are drawn in the polar plot. Thus, a preferred order of the attributes must be defined prior to starting the analysis. The preferred order may be based on the desired sequence in which alternatives must satisfy management goals. For example, it may be desirable to attend to product quality issues before addressing throughput issues. The surface area of the base circle may be interpreted as a measure of the global organizational goal with respect to such performance indicators as available capital, market share, capacity utilization, and so on. Thus, the weighted area of the polyhedron associated with an alternative may be viewed as the degree to which that alternative satisfies organizational goals.

Some of the attributes involved in a selection problem might constitute a combination of quantitative and/or qualitative factors or a combination of objective and/or subjective considerations. The prioritizing of the factors and considerations are typically based on the experience, intuition, and subjective preferences of the decision maker. Goal programming is another technique that can be used to evaluate multiple objectives or criteria in decision problems.

10.3 The analytic hierarchy process

The *analytic hierarchy process* (AHP) is a practical approach to solving complex decision problems involving the pairwise comparisons of alternatives (Saaty, 1980). The technique, popularly known as AHP, has been used extensively in practice to solve many decision problems. AHP enables decision makers to represent the hierarchical interaction of factors, attributes, characteristics, or alternatives in a multifactor decision-making environment. Figure 10.12 presents an example of a decision hierarchy for investment alternatives.

In an AHP hierarchy, the top level reflects the overall objective of the decision problem. The factors or attributes on which the final objective is dependent are listed at intermediate levels in the hierarchy. The lowest level in the hierarchy contains the competing alternatives through which the final objective might be achieved. After the hierarchy has been constructed, the decision maker must undertake a subjective prioritization procedure in order to determine the weight of each element at each level of the hierarchy. Pairwise comparisons are performed at each level to determine the relative importance of each element at that level with respect to each element at the next higher level in the hierarchy. In our example, three alternate investments

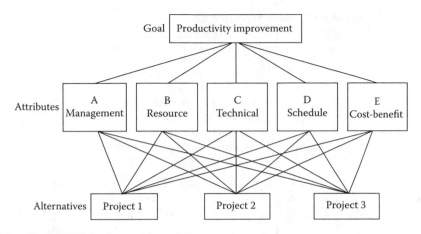

Figure 10.12 AHP for investment alternatives.

are to be considered. The investments are to be compared on the basis of the following five investment attributes:

- Management support
- Resource requirements
- Technical merit
- Schedule effectiveness
- Benefit/cost ratio

The first step in the AHP procedure involves the relative weighting of the attributes with respect to the overall goal. The attributes are compared pairwise with respect to their respective importance to the goal. The pairwise comparison is done through subjective and/or quantitative evaluation by the decision makers. The matrix below shows the general layout for pairwise comparisons.

$$F = \begin{bmatrix} r_{11} & r_{12} & \cdots & r_{1n} \\ r_{21} & r_{22} & \cdots & r_{2n} \\ \cdot & \cdot & \cdots & \cdot \\ \cdot & \cdot & \cdots & \cdot \\ \cdot & \cdot & \cdots & \cdot \\ r_{n1} & r_{n2} & \cdots & r_{nn} \end{bmatrix}$$

where

F = matrix of pairwise comparisons
r_{ij} = relative preference of the decision maker for i to j
$r_{ij} = 1/r_{ji}$
$r_{ij} = 1$

Table 10.8 AHP Weight Scale

Scale	Definition
1	Equal importance
3	Weak importance of one over another
5	Essential importance
7	Demonstrated importance
9	Absolute importance
2, 4, 6, 8	Intermediate weights
Reciprocals	Represented by negative numbers

If $r_{ik}/r_{ij} = r_{jk}$ for all i, j, k, then matrix $\mathbf{F}$ can be said to be perfectly consistent. In other words, the transitivity of the preference orders is preserved. Thus, if Factor A is preferred to Factor B by a scale of 2, and Factor A is preferred to Factor C by a scale of 6, then Factor B will be preferred to Factor C by a factor of 3. That is, A = 2B and A = 6. Then, 2B = 6C (i.e., B = 3C). The weight ratings used by AHP are summarized in Table 10.8.

In practical situations, one cannot expect all pairwise comparison matrices to be perfectly consistent. Thus, a tolerance level for consistency was developed by Saaty (1980). The tolerance level, referred to as the consistency ratio, is acceptable if it is less than 0.10 (10%). If a consistency ratio is greater than 10%, the decision maker has the option of going back to reconstruct the comparison matrix or of proceeding with the analysis, with the recognition, however, that he or she accepts the potential bias that may exist in the final decision. Once the pairwise comparisons matrix is complete, the relative weights of the factors included in the matrix are obtained from the estimate of the maximum eigenvector of the matrix. This is done by the expression below:

$$FW = \lambda_{\max} W$$

where

$\mathbf{F}$ = matrix of pairwise comparisons
$\lambda_{\max}$ = maximum eigenvector of $\mathbf{F}$
$\mathbf{W}$ = vector of relative weights

Example 10.4

For the example in Figure 10.12, Table 10.9 shows the tabulation of the pairwise comparison of the investment attributes.

Each of the attributes listed along the rows of the table is compared against each of the attributes listed in the columns. Each number in the body of the table indicates the degree of preference or importance of one attribute over the other on a scale of 1 to 9. A typical question that may be used to arrive at the relative rating is the following:

Table 10.9 Pairwise Comparisons of Investment Attributes

Attributes	Management	Resource	Technical	Schedule	Benefit/Cost
Management	1	1/3	5	6	5
Resource	2	1	6	7	6
Technical	1/5	1/6	1	3	1
Schedule	1/6	1/7	1/3	1	1/4
Benefit/Cost	1/5	1/6	1	4	1

"With respect to the goal of improving productivity, do you consider investment resource requirements to be more important than technical merit?"

"If so, how much more important is it on a scale of 1 to 9?"

Similar questions are asked iteratively until each attribute has been compared with each of the other attributes. For example, in Table 10.9, Attribute B (resource requirements) is considered to be more important than Attribute C (technical merit) with a degree of 6. In general, the numbers indicating the relative importance of the attributes are based on the following weight scales:

1: Equally important
3: Slightly more important
5: Strongly more important
7: Very strongly more important
9: Absolutely more important

Intermediate ratings are used as appropriate to indicate intermediate levels of importance. If the comparison order is reversed (e.g., B vs. A rather than A vs. B), then the reciprocal of the important rating is entered in the pairwise comparison table. The relative evaluation ratings in the table are converted to the matrix of pairwise comparisons shown in Table 10.10.

The entries in Table 10.10 are normalized to obtain Table 10.11. The normalization is done by dividing each entry in a column by the sum of all the entries in the column. For example, the first cell in Table 10.11 (i.e., 0.219)

Table 10.10 Pairwise Comparisons of Investment Attributes

Attributes	A	B	C	D	E
A	1.000	0.333	5.000	6.000	5.000
B	3.000	1.000	6.000	7.000	6.000
C	0.200	0.167	1.000	3.000	1.000
D	0.167	0.143	0.333	1.000	0.250
E	0.200	0.167	1.000	4.000	1.000
Column Sum	*4.567*	*1.810*	*13.333*	*21.000*	*13.250*

Table 10.11 Normalized Matrix of Pairwise Comparisons

Attributes	A	B	C	D	E	Sum	*Average*
A: Management support	0.219	0.184	0.375	0.286	0.377	1.441	*0.288*
B: Resource requirement	0.656	0.551	0.450	0.333	0.454	2.444	*0.489*
C: Technical merit	0.044	0.094	0.075	0.143	0.075	0.431	*0.086*
D: Schedule effectiveness	0.037	0.077	0.025	0.048	0.019	0.206	*0.041*
E: Benefit/Cost ratio	0.044	0.094	0.075	0.190	0.075	0.478	*0.096*
Column Sum	*1.000*	*1.000*	*1.000*	*1.000*	*1.000*		*1.000*

is obtained by dividing 1.000 by 4.567. Note that the sum of the normalized values in each attribute column is 1.

The last column in Table 10.11 shows the normalized average rating associated with each attribute. This column represents the estimated maximum eigenvector of the matrix of pairwise comparisons. The first entry in the column (0.288) is obtained by dividing 1.441 by 5, which is the number of attributes. The averages represent the relative weights (between 0.0 and 1.0) of the attributes that are being evaluated. The relative weights show that Attribute B (resource requirements) has the highest importance rating of 0.489. Thus, for this example, resource consideration is seen to be the most important factor in the selection of one of the three alternate investments. It should be emphasized that these attribute weights are valid only for the particular goal specified in the AHP model for the problem. If another goal is specified, the attributes would need to be reevaluated with respect to that new goal.

After the relative weights of the attributes are obtained, the next step is to evaluate the alternatives on the basis of the attributes. In this step, relative evaluation rating is obtained for each alternative with respect to each attribute. The procedure for the pairwise comparison of the alternatives is similar to the procedure for comparing the attributes. Table 10.12 presents the tabulation of the pairwise comparisons of the three alternatives with respect to Attribute A (management support).

The table shows that Investment 1 and Investment 3 have the same level of management support. Examples of questions that may be used in obtaining the pairwise ratings of the alternatives are these:

"Is Investment 1 preferable to Investment 2 with respect to management support?"
"What is the level of preference on a scale of 1 to 9?"

Table 10.12 Investment Ratings Based on Management Support

Alternatives	Investment 1	Investment 2	Investment 3
Investment 1	1	1/3	1
Investment 2	3	1	2
Investment 3	1	1/2	1

Table 10.13 Investment Weights Based on Attributes

	Attributes				
	Management Support	Resource Requirements	Technical Merit	Schedule Effectiveness	C/B Ratio
Investment 1	0.21	0.12	0.50	0.63	0.62
Investment 2	0.55	0.55	0.25	0.30	0.24
Investment 3	0.24	0.33	0.25	0.07	0.14

It should be noted that the comparisons shown in Table 10.11 are valid only when management support of the investments is being considered. Separate pairwise comparisons of the investments must be done whenever another attribute is being considered. Consequently, for our example, we would have five separate matrices of pairwise comparisons of the alternatives — one matrix for each attribute. Table 10.12 is the first of the five matrices; the other four are not shown. The normalization of the entries in Table 10.12 yields the following relative weights of the investments with respect to management support: Investment 1 (0.21), Investment 2 (0.55), and Investment 3 (0.24). Table 10.13 shows a summary of the normalized relative ratings of the three investments with respect to each of the five attributes.

The attribute weights shown in Table 10.12 are combined with the weights in Table 10.13 to obtain the overall relative weights of the investments as shown below:

$$\alpha_j = \sum_i \left(w_i k_{ij} \right)$$

where

α_j = *overall* weight for Investment j
w_i = relative weight for attribute i
k_{ij} = rating (local weight) for Investment j with respect to attribute i
$w_i k_{ij}$ = global weight of alternative j with respect to attribute i

Table 10.14 shows the summary of the final AHP analysis for the example. The summary shows that Investment 2 should be selected because it has the highest overall weight of 0.484. AHP can be used to prioritize multiple investments with respect to several objectives. Table 10.15 shows a generic layout for a multiple investments evaluation.

10.4   Investment benchmarking

The techniques presented in the preceding sections can be used for benchmarking investments. For example, to develop a baseline schedule, evidence of successful practices from other investments may be needed. Metrics based

Table 10.14 Summary of AHP Evaluation of Three Investments

	Attributes					
	A $i = 1$	B $i = 1$	C $i = 1$	D $I = 1$	E $i = 1$	
$w_i \Rightarrow$	0.288	0.489	0.086	0.041	0.096	
Investment j			k_{ij}			α_j
Investment 1	0.21	0.12	0.50	0.63	0.62	*0.248*
Investment 2	0.55	0.55	0.25	0.30	0.24	*0.484*
Investment 3	0.24	0.33	0.25	0.07	0.14	*0.268*
Column Sum	*1.000*	*1.000*	*1.000*	*1.000*	*1.000*	*1.000*

Table 10.15 Layout for Multiple Investments Comparison

	Objectives				
	Objective 1	Objective 2	Objective 3	.....	Objective 4
Investment 1	K_{11}	K_{12}	K_{13}		K_{14}
Investment 2	K_{21}	K_{22}	K_{23}		K_{24}
Investment 3	K_{31}	K_{32}	K_{33}		K_{34}
...					
...					
Investment n	K_{n1}	K_{n2}	K_{n3}	.....	K_{nn}

on an organization's most critical investment implementation issues should be developed. *Benchmarking* is a process whereby target performance standards are established based on the best examples available. The objective is to equal or surpass the best example. In its simplest term, benchmarking means learning from and emulating a superior example. The premise of benchmarking is that, if an organization replicates the best quality examples, it will become one of the best in the industry. A major approach of benchmarking is to identify performance gaps between investments. Figure 10.13 shows an example of such a gap. Benchmarking requires that an attempt be made to close the gap by improving the performance of the investment subject.

Benchmarking requires frequent comparison with the target investment. Updates must be obtained from investments already benchmarked, and new investments to be benchmarked must be selected on a periodic basis. Measurement, analysis, feedback, and modification should be incorporated into the performance improvement program. The benchmark–feedback model presented in Figure 10.14 is useful for establishing a continuous drive towards performance benchmarks.

The figure shows the block diagram representation of input–output relationships of the components in a benchmarking environment. In the model, $I(t)$ represents the set of benchmark inputs to the investment subject. The inputs may be in terms of data, information, raw material, technical skill, or

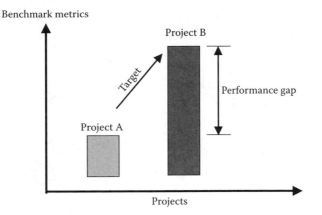

Figure 10.13 Identification of benchmark gaps.

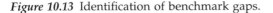

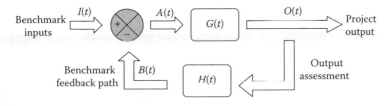

Figure 10.14 Investment benchmark–feedback model.

other basic resources. The index t denotes a time reference. $A(t)$ represents the feedback loop actuator. The actuator facilitates the flow of inputs to the various segments of the investment. $G(t)$ represents the forward transfer function, which coordinates input information and resources to produce the desired output, $O(t)$. $H(t)$ represents the management control process that monitors the status of improvement and generates the appropriate feedback information, $B(t)$, which is routed to the input transfer function. The feedback information is necessary to determine what control actions should be taken at the next improvement phase. The primary responsibility of an economic analyst is to ensure proper forward and backward flow of information concerning the performance of an investment on the basis of the benchmarked inputs.

References

Park, C.S. and Sharp-Bette, G.P., *Advanced Engineering Economics*, Wiley and Sons, New York, 1990.

Troxler, J.W. and Blank, L., A comprehensive methodology for manufacturing system evaluation and comparison, *Journal of Manufacturing Systems*, Vol. 8, No. 3, 176–183, 1989.

Badiru, A.B., *Project Management Tools for Engineering and Management Professionals*, Industrial Engineering and Management Press, Norcross, GA, 1991.

Saaty, T.L., *The Analytic Hierarchy Process*, McGraw-Hill, New York, 1980.

chapter eleven

Budgeting and capital allocation

Budgeting involves sharing limited resources among several project groups or functions contained in a project. A budget *analysis* can serve as any of the following:

- A plan for resources expenditure
- A project selection criterion
- A projection of project policy
- A basis for project control
- A performance measure
- A standardization of resource allocation
- An incentive for improvement

11.1 Top-down budgeting

Top-down budgeting involves collecting data from upper-level sources such as top and middle managers. The figures supplied by the managers may come from their personal judgment, past experience, or past data on similar project activities. The cost estimates are passed on to lower-level managers who then break the estimates down into specific work components within the project. These estimates may, in turn, be given to line managers, supervisors, and lead workers to continue the process until individual activity costs are obtained. Top management provides the global budget, whereas the functional-level worker provides specific budget requirements for project items.

11.2 Bottom-up budgeting

In this method, elemental activities and their schedules, descriptions, and labor skill requirements are used to construct detailed budget requests. Line workers familiar with specific activities are asked to provide cost estimates.

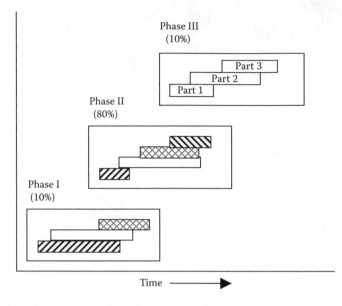

Figure 11.1 Budgeting by project phases.

Estimates are made for each activity in terms of labor time, materials, and machine time. The estimates are then converted to an appropriate cost basis. The dollar estimates are combined into composite budgets at each successive level up the budgeting hierarchy. If estimate discrepancies develop, they can be resolved through the intervention of senior management, middle management, functional managers, the project manager, accountants, or standard cost consultants. Figure 11.1 shows the breakdown of a project into phases and parts in order to facilitate bottom-up budgeting as well as improving both schedule and cost control.

　　Elemental budgets may be developed on the basis of the timed progress of each part of the project. When all the individual estimates are gathered, a composite budget can be developed. Figure 11.2 shows an example of the various components that may be involved in an overall budget. The bar chart appended to a segment of the pie chart indicates the individual cost components making up that particular segment. Such analytical tools as learning-curve analysis, work sampling, and statistical estimation may be employed in the cost estimation and budgeting processes.

11.3　*Mathematical formulation of capital allocation*

Capital rationing involves selecting a combination of projects that will optimize the return on investment (ROI). A mathematical formulation of the capital budgeting problem is presented here:

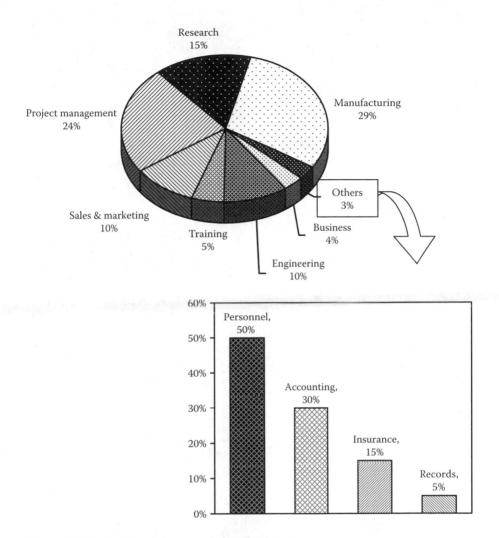

Figure 11.2 Budget breakdown and distribution.

$$\text{Maximize } z = \sum_{i=1}^{n} v_i x_i$$

$$\text{Subject to } \sum_{i=1}^{n} c_i x_i \le B$$

$$x_i = 0, 1; \quad i = 1, \dots, n$$

where

> n = number of projects
> v_i = measure of performance for project i (e.g., present value)
> c_i = cost of project i
> x_i = indicator variable for project i
> B = budget availability level

A solution of this model will indicate what projects should be selected in combination with other projects. The example that follows illustrates a capital rationing problem.

Example 11.1: Capital Rationing Problem

Planning a portfolio of projects is essential in resource-limited projects. The capital-rationing example presented here demonstrates how to determine the optimal combination of project investments so as to maximize total ROI. Suppose a project analyst is given N projects, X_1, X_2, X_3, ... X_N, with the requirement to determine the level of investment in each project so that total investment return is maximized subject to a specified limit on the available budget. The projects are not mutually exclusive.

The investment in each project starts at a base level b_i (i = 1, 2, ..., N) and increases by variable increments k_{ij} (j = 1, 2, 3, ... K_i), where K_i is the number of increments used for project i. Consequently, the level of investment in project X_i is defined as follows:

$$x_i = b_i + \sum_{j=1}^{K_i} k_{ij}$$

where

$$x_i \geq 0, \quad \forall i$$

For most cases, the base investment will be zero. In those cases, we will have $b_i = 0$. In the modeling procedure used for this problem, we have

$$X_i = \begin{cases} 1 & \text{if the investment in project } i \text{ is greater than zero} \\ 0 & \text{otherwise} \end{cases}$$

and

$$Y_{ij} = \begin{cases} 1 & \text{if the } j\text{th increment of alternative } i \text{ is used} \\ 0 & \text{otherwise} \end{cases}$$

The variable x_i is the actual level of investment in project i, whereas X_i is an indicator variable indicating whether or not project i is one of the projects selected for investment. Similarly, k_{ij} is the actual magnitude of the jth increment, whereas Y_{ij} is an indicator variable that indicates whether or not the jth increment is used for project i. The maximum possible investment in each project is defined as M_i, such that

$$b_i \le x_i \le M_i$$

There is a specified limit, B, on the total budget available to invest, such that

$$\sum_i x_i \le B$$

There is a known relationship between the level of investment, x_i, in each project and the expected return, $R(x_i)$. This relationship will be referred to as the *utility function*, f(.) for the project. The utility function may be developed through historical data, regression analysis, and forecasting models. For a given project, the utility function is used to determine the expected return, $R(x_i)$, for a specified level of investment in that project. That is,

$$R(x_i) = f(x_i)$$

$$= \sum_{j=1}^{K_i} r_{ij} Y_{ij}$$

where r_{ij} is the incremental return obtained when the investment in project i is increased by k_{ij}. If the incremental return decreases as the level of investment increases, the utility function will be concave. In that case, we will have the following relationship:

$$r_{ij} \ge r_{ij+1} \quad \text{or} \quad r_{ij} - r_{ij+1} \ge 0$$

Thus,

$$Y_{ij} \ge Y_{ij+1} \quad \text{or} \quad Y_{ij} - Y_{ij+1} \ge 0$$

so that only the first n increments (j = 1, 2, ..., n) that produce the highest returns are used for project i. Figure 11.3 shows an example of a concave investment utility function.

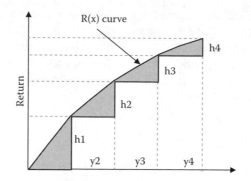

Figure 11.3 Utility curve for investment yield.

If the incremental returns do not define a concave function, $f(x_i)$, then one has to introduce the inequality constraints presented previously into the optimization model. Otherwise, the inequality constraints may be left out of the model because the first inequality, $Y_{ij} \geq Y_{ij+1}$, is always implicitly satisfied for concave functions. Our objective is to maximize the total return. That is,

$$\text{Maximize} \quad Z = \sum_i \sum_j r_{ij} Y_{ij}$$

subject to the following constraints:

$$x_i = b_i + \sum_j k_{ij} Y_{ij} \quad \forall i$$

$$b_i \leq x_i \leq M_i \quad \forall i$$

$$Y_{ij} \geq Y_{ij+1} \quad \forall i,j$$

$$\sum_i x_i \leq B$$

$$x_i \geq 0 \quad \forall i$$

$$Y_{ij} = 0 \text{ or } 1 \quad \forall i,j$$

Now, suppose we are given four projects (i.e., $N = 4$) and a budget limit of $10 million. The respective investments and returns are shown in Table 11.1, Table 11.2, Table 11.3, and Table 11.4. All of the values are in millions of dollars. For example, in Table 11.1, if an incremental investment of $0.20 million from stage 2 to stage 3 is made in Project 1, the expected incremental return from the project will be $0.30 million. Thus, a total investment of $1.20 million in Project 1 will yield a total return of $1.90 million. The question addressed by the optimization model is to determine how many investment increments should be used for each project — that is, when should we stop increasing the investments in a given project? Obviously, for a single project, we would continue to invest as long as the incremental returns are larger than the incremental investments. However, for multiple projects, investment interactions complicate the decision so that investment in one project cannot be independent of the other projects. The Linear Programming (LP) model of the capital-rationing example was solved with LINDO software. The solution indicates the following values for Y_{ij}.

Project 1

$$Y11 = 1, Y12 = 1, Y13 = 1, Y14 = 0, Y15 = 0$$

Thus, the investment in project 1 is $X_1 = \$1.20$ million. The corresponding return is $\$1.90$ million.

Table 11.1 Investment Data for Project 1 for Capital Rationing

Stage (j)	y_{ij} Incremental Investment	x_1 Level of Investment	r_{ij} Incremental Return	$R(x_1)$ Total Return
0	—	0	—	0
1	0.80	0.80	1.40	1.40
2	0.20	1.00	0.20	1.60
3	0.20	1.20	0.30	1.90
4	0.20	1.40	0.10	2.00
5	0.20	1.60	0.10	2.10

Project 2

$$Y21 = 1, Y22 = 1, Y23 = 1, Y24 = 1, Y25 = 0, Y26 = 0, Y27 = 0$$

Thus, the investment in project 2 is $X_2 = \$3.80$ million. The corresponding return is $\$6.80$ million.

Table 11.2 Investment Data for Project 2 for Capital Rationing

Stage (j)	y_{2j} Incremental Investment	x_2 Level of Investment	r_{2j} Incremental Return	$R(x_2)$ Total Return
0	—	0	—	0
1	3.20	3.20	6.00	6.00
2	0.20	3.40	0.30	6.30
3	0.20	3.60	0.30	6.60
4	0.20	3.80	0.20	6.80
5	0.20	4.00	0.10	6.90
6	0.20	4.20	0.05	6.95
7	0.20	4.40	0.05	7.00

Project 3

Y31 = 1, Y32 = 1, Y33 = 1, Y34 = 1, Y35 = 0, Y36 = 0 , Y37 = 0

Thus, the investment in project 3 is $X_3 = \$2.60$ million. The corresponding return is $5.90 million.

Table 11.3 Investment Data for Project 3 for Capital Rationing

Stage (j)	y_{3j} Incremental Investment	x_3 Level of Investment	r_{3j} Incremental Return	$R(x_3)$ Total Return
0	0	—	—	0
1	2.00	2.00	4.90	4.90
2	0.20	2.20	0.30	5.20
3	0.20	2.40	0.40	5.60
4	0.20	2.60	0.30	5.90
5	0.20	2.80	0.20	6.10
6	0.20	3.00	0.10	6.20
7	0.20	3.20	0.10	6.30
8	0.20	3.40	0.10	6.40

Project 4

Y41 = 1, Y42 = 1, Y43 = 1

Thus, the investment in project 4 is $X_4 = \$2.35$ million. The corresponding return is $3.70 million.

Table 11.4 Investment Data for Project 4 for Capital Rationing

Stage (j)	y_{4j} Incremental Investment	x_4 Level of Investment	r_{4j} Incremental Return	$R(x_4)$ Total Return
0	—	0	—	0
1	1.95	1.95	3.00	3.00
2	0.20	2.15	0.50	3.50
3	0.20	2.35	0.20	3.70
4	0.20	2.55	0.10	3.80
5	0.20	2.75	0.05	3.85
6	0.20	2.95	0.15	4.00
7	0.20	3.15	0.00	4.00

The total investment in all four projects is $9,950,000. Thus, the optimal solution indicates that not all of the $10,000,000 available should be invested. The expected return from the total investment is $18,300,000. This translates into an 83.92% return on investment. Figure 11.4 presents histograms of the investments and the returns for the four projects. The individual returns on investment from the projects are shown graphically in Figure 11.5.

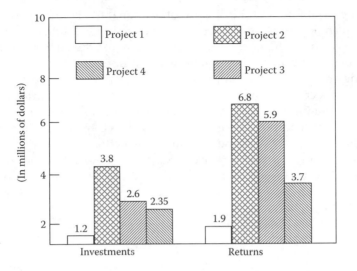

Figure 11.4 Histogram of capital-rationing example.

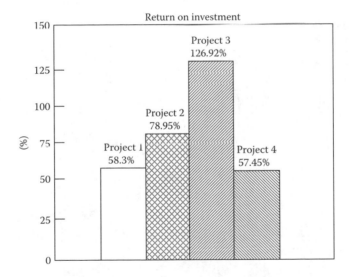

Figure 11.5 Histogram of ROI.

The optimal solution indicates an unusually large return on total invest-ment. In a practical setting, expectations may need to be scaled down to fit the realities of the project environment. Not all optimization results will be directly applicable to real situations. Possible extensions of the preceding model of capital rationing include the incorporation of risk and time value of money into the solution procedure. Risk analysis would be relevant, particularly for cases where the levels of returns for the various levels of investment are not known with certainty. The incorporation of time value of money would be useful if the investment analysis is to be performed for a given planning horizon. For example, we might need to make investment decisions to cover the next 5 years rather than just the current time.

As mentioned at the beginning of this chapter, budgeting and capital-rationing processes convey important messages about the investment potential of a project. The techniques presented in this chapter offer additional tools for assessing how and where limited resources should be directed. The outputs of the computational analysis can serve as a plan for resources expenditure, a project selection criterion, a projection of project policy, a basis for project control, a performance measure, a standardization of resource allocation, and an incentive for improvement of operational practices.

chapter twelve

The ENGINEA software: a tool for economic evaluation

This chapter presents some of the software tools used in engineering economic evaluation. The use and importance of software tools in engineering economic evaluation, especially spreadsheet functions, has been emphasized in the literature (Omitaomu et al., 2005). Spreadsheets provide rapid solutions, and the results of the analyses can be readily saved for easy presentation and for future reference.

12.1 The ENGINEA software

The *ENGINEA* software is a MDI (multiple-document interface) decision-based application, developed with Visual Basic software, to help students and engineers perform several types of economic analysis. Because it is a MDI software, users can work with several documents (cases) simultaneously. The *ENGINEA* software was developed to target three major audiences: undergraduate students, graduate students, and instructors teaching both engineering economic analysis and financial management courses. In addition, the software should also be useful to students in upper undergraduate and graduate classes, to engineering practitioners in the industry, to financial analysts, to recruiters for promoting the advancement of economic analysis, and, finally, for the purpose of training engineers. Therefore, the objectives of the *ENGINEA* software include the following:

- To introduce to engineering economic analysis and financial-management students a creative and hands-on approach for solving complex economic analysis problems
- To provide a tool specifically developed for solving engineering economic analysis problems

- To demonstrate how engineers can use computer tools for developing decision-support modules
- To enhance the teaching and dissemination of techniques for solving engineering economic analysis problems

Useful functionality and user friendliness were two major factors considered when the *ENGINEA* application was being designed. Because hard-copy results are valuable for future reference, the ability to print the inputs and the results is available for all modules. The outputs can be saved to a file at any time for future reference. Example files are provided in the software installation, which can be loaded for the user's reference. A standard menu bar allows the user to open the modules, to save, and to print the output. In addition, a graphical toolbar containing easily recognizable icons is available to enable the user to perform the same functions available in the menu bar. The application is stable in the current release. No errors in the application are known at this time. All invalid user inputs are handled appropriately. As many components in the application desktop as desired can be simultaneously opened, giving the user the ability to easily compare results.

12.1.1   Instructional design

The *ENGINEA* application can be used to enhance user knowledge of cash-flow analysis, benefit/cost ratio, rate of return, replacement analysis, and depreciation analysis problems in engineering economy and financial management courses. These analyses are the core areas of the *ENGINEA* software. A factor interest calculator is also available for finding factor values more rapidly and easily.

12.1.2   Cash-flow analysis

As cash-flow values for each period are entered, it is easy to see how the cash-flow diagram is constructed. This is one of the unique features of the software. In addition, a user can see how his or her input affects the present worth (PW), the annual worth (AW), and the future worth (FW) of the cash-flow profile. The interest rate can also be changed at any time, so the user can learn how changing interest rates affect the present worth, the annual worth, and the future worth values. A screenshot of a cash-flow analysis example is shown in Figure 12.1.

Figure 12.1 shows the cash-flow diagram, the PW, FW, AW, and the interest rate. The user is presented with a large scrollable area for the cash-flow diagram because the diagram is a vital portion of most cash-flow problems. The user can add, update, and remove cash-flow values using the appropriate buttons. The user also has the ability to move to any period using the left and right arrow buttons, and the currently selected cash-flow

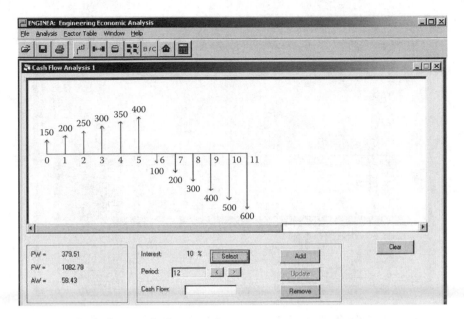

Figure 12.1 Cash-flow analysis screenshot.

period is highlighted in the diagram. The maximum number of periods allowed is 32. The interest rate can also be updated for sensitivity analysis. In all cases, the PW, AW, and FW are updated automatically. As stated previously, this profile can be saved for future reference or usage, and it can also be printed for reporting purposes.

12.1.3 Replacement analysis

In the Replacement Analysis module, the Equivalent Uniform Annual Cost (EUAC), the Economic Service Life, and the Minimum EUAC are dynamically calculated as the user enters Market Value (MV), Annual Expenses (AE), and Total Marginal Cost (TMC) values of either the challenger or the defender. The users can see how the values they enter affect the minimum EUAC and ultimately the result of the analysis. When the user chooses, he or she may view the analysis result, which will aid in a decision either to keep the defender or to replace the defender with the challenger, as shown in Figure 12.2.

Replacement analysis can be done by inputting either both the MV and AE, as shown in Figure 12.2a, or the TMC, as shown in Figure 12.2b, for both the challenger and the defender. Figure 12.2a is a replacement analysis table for the defender; whereas Figure 12.2b is a replacement analysis table for the challenger. The result of the decision about the replacement analysis generated at the click of the result button is also shown. Replacement analysis is

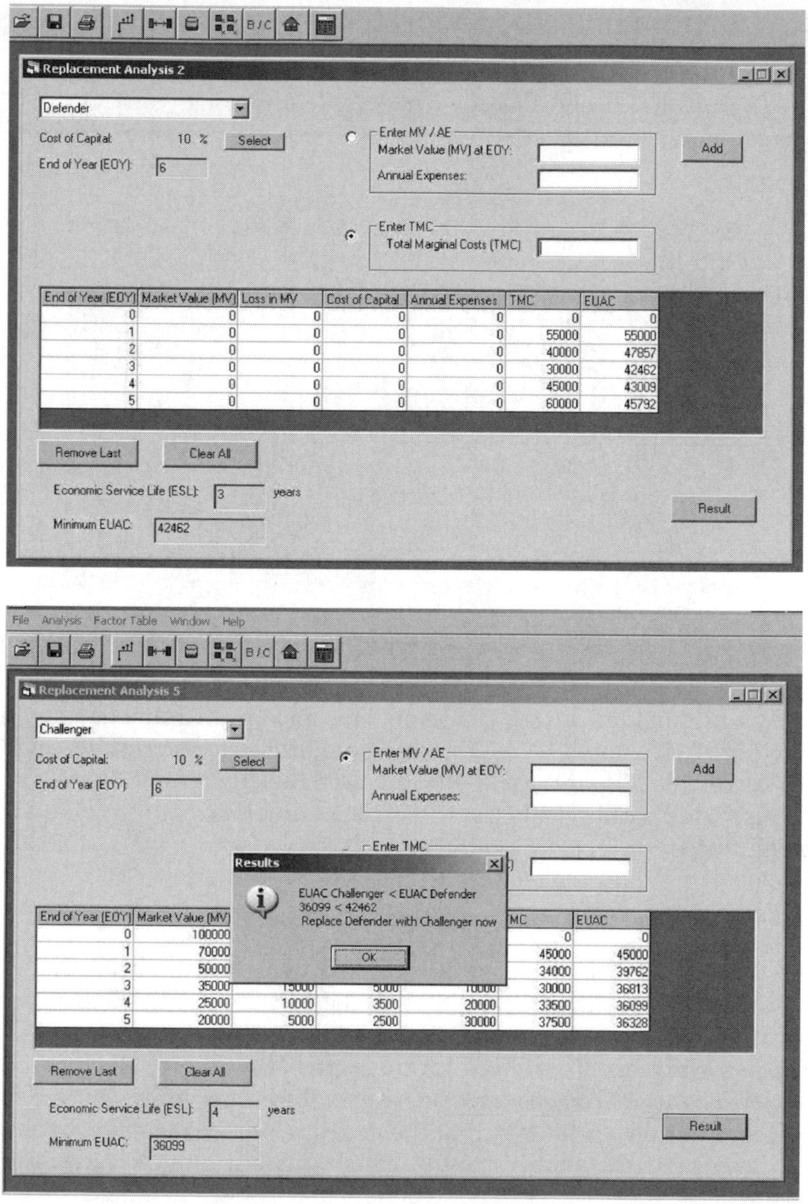

Figure 12.2 Replacement analysis platform screenshots.

computationally intensive; therefore, students can easily make mistakes. This software can help students check their results easily. For this module, the challenger and defender use the same input screen; therefore, a drop-down list is provided to select either the challenger or the defender.

12.1.4 Depreciation analysis

This module (Figure 12.3) allows the user to manipulate the initial cost, salvage value, and useful life of a depreciable asset using any of the major depreciation methods. The depreciation methods supported by the software are the following: straight-line (SL), sum-of-years digit (SYD), declining balance (DB) (with the opportunity to specify the depreciation rate), double declining balance (DDB), and modified accelerated cost recovery system (MACRS).

Figures 12.3a and 12.3b are tabular representations of the depreciation amounts and the corresponding book values for SL and MACRS depreciation methods, respectively. The depreciation analysis component is structured so that the user can employ any of the depreciation methods by entering values for the asset initial cost, salvage value, and useful life. The salvage value is only active if the SL or SYD method is selected, because those are the only two methods that use salvage value in their equations. The application will also display a friendly error message to correct an undefined asset recovery period if MACRS is selected. For example, if a useful life of 6 years was input, the application will display an error message that such a recovery period is not applicable for the chosen method.

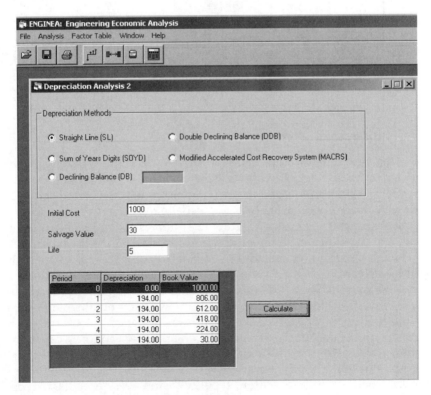

Figure 12.3 Depreciation analysis screenshots.

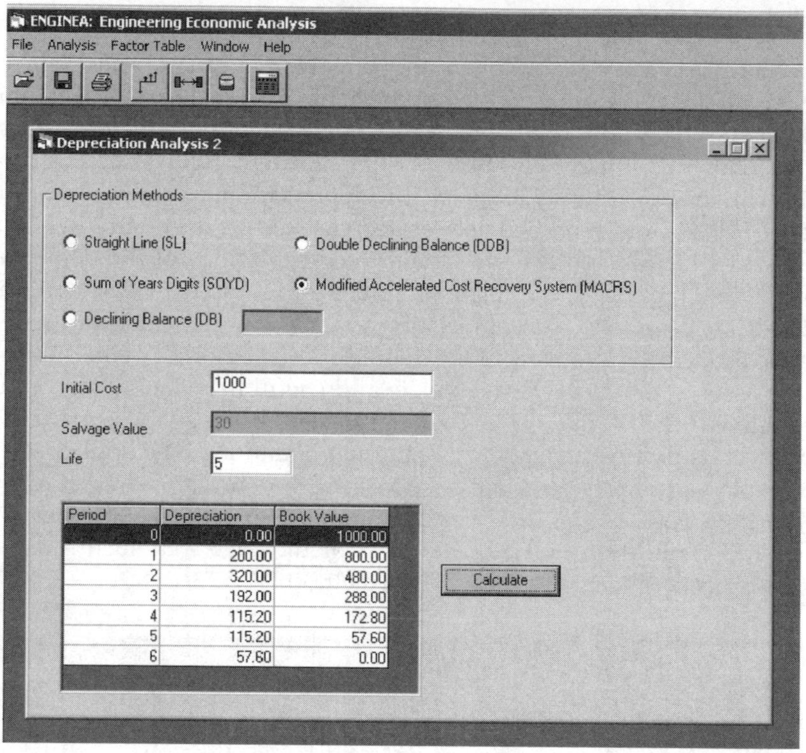

Figure **12.3** (continued).

12.1.5   *Interest calculator*

One of the most difficult steps in economic analysis is interpolating between tabulated interest rates. In cases where such interpolation can be easily achieved, the assumption of a linear relationship between tabulated values is not realistic in economic analysis and usually results in errors of up to 5%. The interest factor calculator (Figure 12.4) allows users to obtain interest factors for various interest values and periods without interpolation.

The interest factor calculator screenshot in Figure 12.4 gives the values for three factors: P/F, P/A, and A/G. Ordinarily, the interest rates and number of periods shown in this screenshot cannot be easily determined using tabulated factor tables, but they are now easily obtainable using the *ENGINEA* software. The interest calculator was designed to look and operate like a standard calculator, except for the engineering economy capabilities. To use the calculator, the user must first select a function before entering the interest value and periods. For example, to compute (P/F, 14.25%, 23), the user will click on the P/F button, type 14.25 and press enter, and type 23 and press enter again for the result of 0.0572. The user has the ability to scroll through the output to see past results.

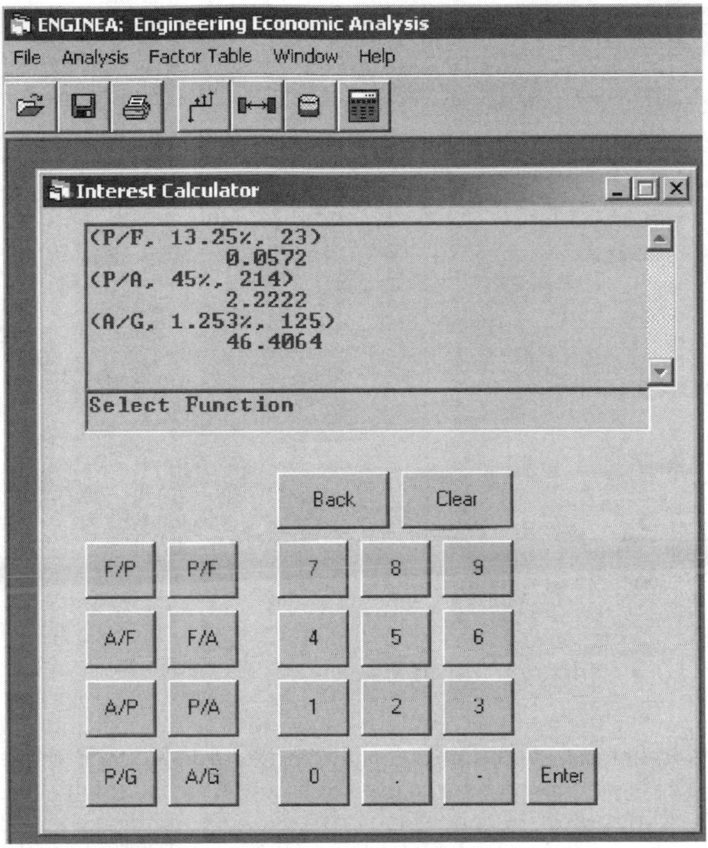

Figure 12.4 Interest calculator module.

12.1.6 Benefit/cost (B/C) ratio analysis

We usually teach our students to use incremental analysis when evaluating projects with B/C ratio or internal rate of return (ROR) analyses. The *ENGINEA* software uses incremental analyses for the B/C ratio and the internal rate of return. The module uses annual cost and benefits; therefore, if data are not available in annual forms, the cash-flow analysis module, described previously, can be used to get the appropriate annual cost and benefits because this is basically a cash-flow problem. The software evaluates the alternatives and displays the decision at each stage of the process, as shown in Figure 12.5. To use this module, the name of the alternative, the annual benefits, and the annual cost must be given. The alternatives must be ranked in order of increasing initial cost. The user clicks on the "Add" button to store data about the alternative. This step is repeated for other alternatives. Once the data about all the alternatives have been given, the "Results" button is clicked for the results of the incremental B/C ratio analysis.

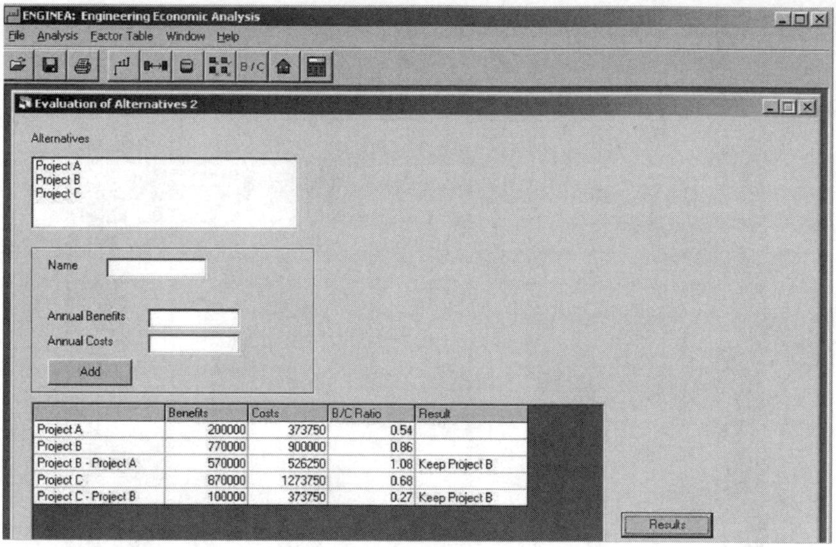

Figure 12.5 Benefit/cost ratio analysis module.

12.1.7 Loan/mortgage analysis

The loan/mortgage analysis module computes the amortization of capital using the initial loan amount, the nominal interest rate, and the length of the loan. The legend to the column titles in Figure 12.6a is shown at the bottom of the analysis table. This analysis is based on the amortization of capital equations as presented by Badiru (1996), and this module calculates the unpaid balance, the monthly payment, the total installment payment, the cumulative equity, and the interest charge per period. In addition, when the graph button is clicked, the module calculates the break-even point for the loan being analyzed. It displays a plot of unpaid balance vs. cumulative equity and shows the break-even point. As discussed previously, the break-even point is the period at which the unpaid balance on a loan equals the cumulative equity on the loan (Badiru, 1996). This point is very important for loan/mortgage analysis. Figure 12.6a presents a portion of an output for a $500,000 loan at 10% for 15 years. The results show a monthly payment of $5,373.03 (the $A(t)$ column) and after the 15th month's payment, the balance on the loan is $480,809.96 (the $U(t)$ column). The equity portion of payment for that period is $1,354.98 (the $E(t)$ column); the interest portion of payment for that period is $4,018.04 (the $I(t)$ column); the total payment made on the loan up to the 15th month is $80,595.38 (the $C(t)$ column), of which only $19,190.04 (the $S(t)$ column) is the total equity to that date. The break-even point for this loan arrangement is 120 months (10 years), as is shown in Figure 12.6b.

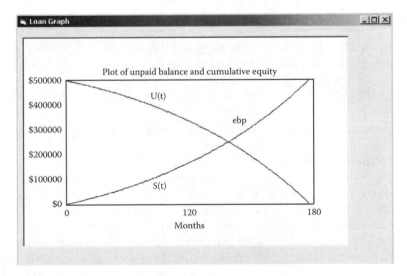

t	U(t)	A(t)	E(t)	I(t)	C(t)	Q(t)	S(t)	f(t)	F(t)	H(t)
1	498793.64	5373.03	1206.36	4166.67	5373.03	4166.67	1206.36	77.55	77.55	22.45
2	497577.23	5373.03	1216.41	4156.61	10746.05	8323.28	2422.77	77.36	77.45	22.64
3	496350.68	5373.03	1226.55	4146.48	16119.08	12469.76	3649.32	77.17	77.36	22.83
4	495113.91	5373.03	1236.77	4136.26	21492.10	16606.01	4886.09	76.98	77.27	23.02
5	493866.84	5373.03	1247.08	4125.95	26865.13	20731.96	6133.17	76.79	77.17	23.21
6	492609.37	5373.03	1257.47	4115.56	32238.15	24847.52	7390.63	76.60	77.07	23.40
7	491341.42	5373.03	1267.95	4105.08	37611.18	28952.60	8658.58	76.40	76.98	23.60
8	490062.91	5373.03	1278.51	4094.51	42984.20	33047.11	9937.10	76.21	76.88	23.80
9	488773.74	5373.03	1289.17	4083.86	48357.23	37130.97	11226.26	76.01	76.78	23.99
10	487473.83	5373.03	1299.91	4073.11	53730.26	41204.08	12526.17	75.81	76.69	24.19
11	486163.08	5373.03	1310.74	4062.28	59103.28	45266.36	13836.92	75.61	76.59	24.39
12	484841.42	5373.03	1321.67	4051.36	64476.31	49317.72	15158.59	75.40	76.49	24.60
13	483508.74	5373.03	1332.68	4040.35	69849.33	53358.07	16491.27	75.20	76.39	24.80
14	482164.95	5373.03	1343.79	4029.24	75222.36	57387.31	17835.05	74.99	76.29	25.01
15	480809.96	5373.03	1354.98	4018.04	80595.38	61405.35	19190.04	74.78	76.19	25.22
16	479443.69	5373.03	1366.28	4006.75	85968.41	65412.10	20556.31	74.57	76.09	25.43

t   Month   E(t):   Equity Portion of the Payment   S(t):   Total Equity to Date
U(t):   Unpaid Balance   I(t):   Interest Charge Contained in the Payment   f(t):   Percentage of Interest Charge
A(t):   Monthly Payment   C(t):   Total Payment to Date   F(t):   Percentage of Cumulative Interest Charge

Figure 12.6 Loan/mortgage analysis screenshots.

12.1.8   Rate-of-return (ROR) analysis

The last module is the ROR analysis module. This module also uses an incremental method to select the best alternatives. It utilizes cash-flow information and the given minimum attractive rate of return (MARR) to compute

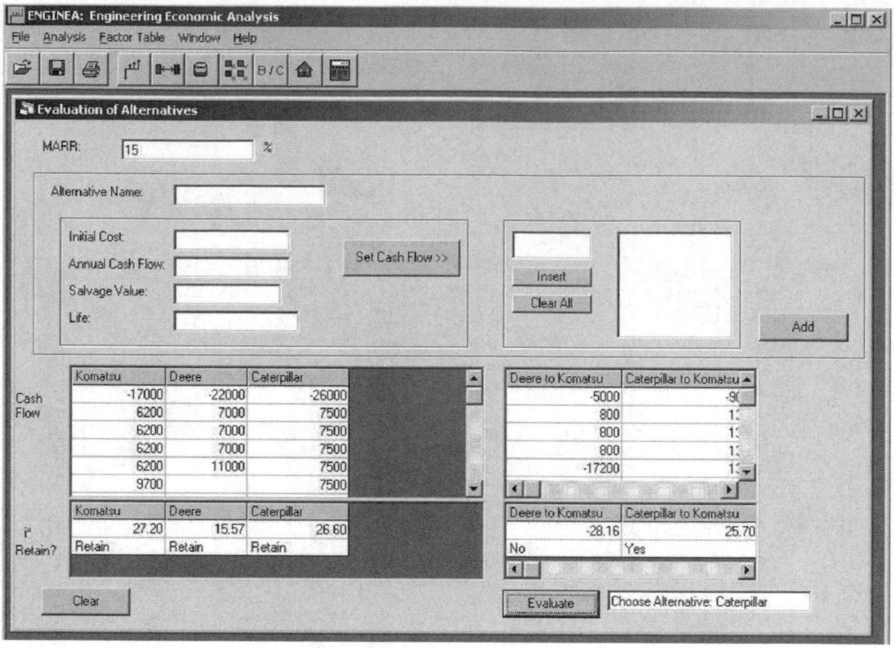

Figure 12.7 ROR analysis module output.

the final results. In the example presented in Figure 12.7, there are three mutually exclusive alternatives, and the module computes their respective RORs and finds the three alternatives to be viable; therefore, the incremental method is used to break the tie. A portion of the output results is shown in Figure 12.7.

To use this module, the MARR for the analysis is stated. Then, for each alternative, the name, the initial cost, the annual cash flow, the salvage value, and the useful life are given. The alternatives must be ranked in order of increasing initial cost. Then the user clicks on the "Set Cash Flow" button to store the data for that alternative, and on the "Add" button to generate the ROR for that alternative. The "Insert" button can be used to add more cash flows after the "Set Cash Flow" button has been clicked or when the "Clear All" button has been used to clear the stored data. This step is repeated for other alternatives. If the computed ROR for each alternative is greater than the given MARR, the alternative is retained in the analysis; otherwise, it is eliminated. If there is more than one alternative with an ROR greater than the MARR, the incremental method is applied. To initialize the incremental method process, the user clicks on the "Evaluate" button, and the analysis is carried out automatically, leaving out all alternatives with an ROR less than the MARR. The final result of the incremental method is shown in the window beside the "Evaluate" button.

12.2 The Excel spreadsheet functions

Some of the commonly used Excel functions in economic evaluation are discussed in the following sections.

12.2.1 DB (declining balance)

- DB (cost, salvage, life, period, month)
- Calculates the depreciation amount for an asset using the declining balance method
    - Cost: First cost or basis of the asset
    - Salvage: Salvage value
    - Life: Recovery period
    - Period: The year for which the depreciation is to be calculated
    - Month (optional): This is the number of months in the first year; a full year (12 months) is assumed for the first year if it is omitted

12.2.2 DDB (double declining balance)

- DDB (cost, salvage, life, period, factor)
- Calculates the depreciation amount for an asset using the double declining balance method
    - Cost: First cost or basis of the asset
    - Salvage: Salvage value
    - Life: Recovery period
    - Period: The year for which the depreciation is to be calculated
    - Factor (optional): Enter 1.5 for 150% declining balance, and so on; the function will use 2.0 for 200% declining balance if omitted

12.2.3 FV (future value)

- FV (rate, nper, pmt, pv, type)
- Calculates the future worth for a periodic payment at a specific interest rate
    - Rate: Interest rate per compounding period
    - Nper: Number of compounding periods
    - Pmt: Constant payment amount
    - Pv: The present value amount; the function will assume that pv is zero if omitted
    - Type (optional): Either 0 or 1; a 0 represents an end-of-the-period payment, and 1 represents a beginning-of-the-period payment; if omitted, 0 is assumed

12.2.4   IPMT (interest payment)

- IPMT (rate, per, nper, pv, fv, type)
- Calculates the interest accrued for a given period based on a constant periodic payment and interest rate
  - Rate: Interest rate per compounding period
  - Per: Period for which interest is to be calculated
  - Nper: Number of compounding periods
  - Pv: The present value amount; the function will assume that pv is zero if omitted
  - Fv: The future value (a cash balance after the last payment is made); if omitted, the function will assume it to be 0.
  - Type (optional): Either 0 or 1; a 0 represents end-of-the-period payment, and 1 represents a beginning-of-the-period payment; if omitted, 0 is assumed

12.2.5   IRR (internal rate of return)

- IRR (values, guess)
- Calculates the internal rate of return between −100% and infinity for a series of cash flows at regular periods
  - Values: A set of numbers in a spreadsheet row or column for which the rate of return will be calculated; there must be at least one positive (cash inflow) and one negative (cash outflow) number
  - Guess (optional): Guess a rate of return to reduce the number of iterations; change the guess if #NUM! error appears

12.2.6   MIRR (modified internal rate of return)

- MIRR (values, finance_rate, reinvest_rate)
- Calculates the MIRR for a series of cash flows and reinvestment of income and interest at a stated rate
  - Values: A set of numbers in a spreadsheet row or column for which the ROR will be calculated; there must be at least one positive (cash inflow) and one negative (cash outflow) number
  - Finance_rate: Interest rate of money used in the cash flows
  - Reinvest_rate: Interest rate for reinvestment on positive cash flows

12.2.7   NPER (number of periods)

- NPER (rate, pmt, pv, fv, type)
- Calculates the number of periods for the present worth of an investment to equal the future value specified
  - Rate: Interest rate per compounding period
  - Pmt: Amount paid during each compounding period

- Pv: Present values
- Fv (optional): The future value (a cash balance after the last payment is made); if omitted, the function will assume it to be 0
- Type (optional): Either 0 or 1; a 0 represents the end-of-the-period payment, and 1 represents the beginning-of-the-period payment; if omitted, 0 is assumed

12.2.8 NPV (net present value)

- NPV (rate, series)
- Calculates the net present worth of a series of future cash flows at a particular interest rate
  - Rate: Interest rate per compounding period
  - Series: Series of inflow and outflow set up in a range of cells in the spreadsheet

12.2.9 PMT (payments)

- PMT (rate, nper, pv, fv, type)
- Calculates equivalent periodic amounts based on present worth and or future worth at a stated interest rate
  - Rate: Interest rate per compounding period
  - Nper: Number of compounding periods
  - Pv: The present value amount. The function will assume that pv is zero if omitted
  - Fv: The future value (a cash balance after the last payment is made); if omitted, the function will assume it to be 0
  - Type (optional): Either 0 or 1; a 0 represents the end-of-the-period payment, and 1 represents the beginning-of-the-period payment; if omitted, 0 is assumed.

12.2.10 PPMT (principal payment)

- PPMT (rate, per, nper, pv, fv, type)
- Calculates the payment on the principal based on uniform payments at a stated interest rate
  - Rate: Interest rate per compounding period
  - Per: Period for which interest is to be calculated
  - Nper: Number of compounding periods
  - Pv: The present value amount; the function will assume that pv is zero if omitted
  - Fv: The future value (a cash balance after the last payment is made); if omitted, the function will assume it to be 0

- Type (optional): Either 0 or 1; a 0 represents the end-of-the-period payment, and 1 represents the beginning-of-the-period payment; if omitted, 0 is assumed

12.2.11   PV (present value)

- PV (rate, nper, pmt, fv, type)
- Calculates the present worth of a future series of equal cash flows and a single lump sum in the last period at a stated interest rate
  - Rate: Interest rate per compounding period
  - Nper: Number of compounding periods
  - Pmt: Cash flow at regular intervals; inflows are positive and outflows are negative
  - Fv: The future value (a cash balance after the last payment is made); if omitted, the function will assume it to be 0
  - Type (optional): Either 0 or 1; a 0 represents the end-of-the-period payment, and 1 represents the beginning-of-the-period payment; if omitted, 0 is assumed

12.2.12   RATE (interest rate)

- RATE (nper, pmt, pv, fv, type, guess)
- Calculates the interest rate per compounding period for a series of payments or incomes
  - Nper: Number of compounding periods
  - Pmt: Cash flow at regular intervals; inflows are positive and outflows are negative
  - Pv: The present value amount; the function will assume that pv is zero if omitted
  - Fv: The future value (a cash balance after the last payment is made); if omitted, the function will assume it to be 0
  - Type (optional): Either 0 or 1; a 0 represents the end-of-the-period payment, and 1 represents beginning-of-the-period payment; if omitted, 0 is assumed
  - Guess (optional): Guess a rate of return to reduce the number of iterations; change the guess if #NUM! error appears

12.2.13   SLN (straight-line depreciation)

- SLN (cost, salvage, life)
- Calculates the straight-line depreciation of an asset for a given year
  - Cost: First cost or basis of the asset
  - Salvage: Salvage value
  - Life: Recovery period

12.2.14   SYD (sum-of-year digits depreciation)

- SYD (cost, salvage, life, period)
- Calculates the sum-of-year digits depreciation of an asset for a given year
  - Cost: First cost or basis of the asset
  - Salvage: Salvage value
  - Life: Recovery period
  - Period: The year for which the depreciation is to be calculated

12.2.15   VDB (variable declining balance)

- VDB (cost, salvage, life, start_period, end_period, factor, no_switch)
- Calculates the depreciation schedule using the declining balance method with a switch to straight-line depreciation in the year in which straight line has a larger depreciation amount; this function can be used for MACRS depreciation schedule computations
  - Cost: First cost or basis of the asset
  - Salvage: Salvage value
  - Life: Recovery period
  - Start_period: First period for depreciation to be calculated
  - End_period: Last period for depreciation to be calculated
  - Factor (optional): Enter 1.5 for a 150% declining balance and so on; the function will use 2.0 for a 200% declining balance if omitted
  - No_switch (optional): If omitted or entered as FALSE, the function will switch from DB or DDB to SLN depreciation when the latter is greater than DB depreciation; if entered as TRUE, the function will not switch to SLN depreciation at any time during the depreciation life

Reference

1. Omitaomu, O.A., Smith, L., and Badiru, A.B., The ENGINeering Economic Analysis (ENGINEA) software: enhancing teaching and application of economic analysis techniques, *Computers in Education*, Vol.15, No. 4, 32–38, 2005.
2. Badiru, Adedeji B., *Project Management and High Technology Operations, 2nd ed.*, John Wiley & Sons, New York, 1996.

chapter thirteen

Cost benchmarking case study*

13.1 Abstract

This case study presents an integrated methodology as a best-practices approach that combines analytical and graphical techniques for cost comparison and benchmarking in new construction projects. The methodology was developed on the basis of a specific scenario of construction projects at a government facility, and it is applicable to both government and commercial construction projects. The methodology determines a set of cost drivers (factors) influencing the cost differences between two comparable projects. The cost difference is then distributed over the cost drivers using an iterative graphical approach. At that point, the area of potential cost improvements can be easily identified.

13.2 Introduction

Cost effectiveness is a major focus in construction project management. Most commercial construction projects take advantage of the best practices available in the industry as a way to achieve a profitable bottom line. Government-funded projects account for a large percentage of construction activities in the U.S., both in terms of capital investment as well as expected impacts. Unfortunately, many government projects are not known for embracing construction industry best practices. Consequently, government projects are notorious for waste and ineffective utilization of resources. Both human capital and material resources are often not efficiently managed or deployed in government construction projects. We would argue that it is imperative that proven best practices be followed in making projects more cost effective.

* Adapted from Badiru, Adedeji B., Delgado, Vincent, Ehresma-Gunter, Jamie, Omitaomu, Olufemi A., Nsofor, Godswill, and Saripali, Sirisha (2004). Graphical and analytical methodology for cost benchmarking for new construction projects. Presented at *13th International Conference on Management of Technology* (*IAMOT* 2004), Washington, DC. April 3–7.

A wave of new construction projects at Oak Ridge National Laboratory (ORNL) in Oak Ridge, TN, created an opportunity to develop new approaches for the cost benchmarking of government construction projects. The laboratory is managed by a consortium commonly referred to as UT–Battelle and consisting of the University of Tennessee and Battelle Corporation. UT–Battelle commissioned the Department of Industrial and Information Engineering (IIE) at the University of Tennessee to examine three different facility construction projects at ORNL's new East Campus Development (ECD), to compare their costs, and to document benchmarks; UT–Battelle also charged IIE with identifying the advantages and disadvantages in the projects' financing and construction methods, and to perform a related analysis of methods used in "best commercial practices." This effort will provide ORNL with more alternatives for the execution of future projects with improved financing and reduced project cost. The three-facility construction projects are listed here and are major constituents of ORNL's Facility Revitalization Project (FRP).

> **Research Center Complex (RCC) Project:** A fast-track, privately funded commercial development consisting of the following facilities:
> - Computational Science Building (CSB)
> - Research Office Building (ROB)
> - Engineering Technology Facility (ETF)
>
> **Joint Institute for Computational Sciences/Oak Ridge Center for Advanced Studies (JICS/ORCAS):** A project funded by the State of Tennessee using its standard process for construction projects: design–bid–build.
>
> **Research Support Center (RSC):** A Department of Energy (DOE)-funded line-item project following a traditional DOE approach.

The three facilities are certified according to Leadership in Energy and Environmental Design (LEED). Because these projects were initiated simultaneously and are being constructed using different methods of execution, they provide a unique benchmarking opportunity. This study examines the RCC project, the JICS/ORCAS project, and the RSC project in terms of their hard and soft costs and various drivers in order to ascertain opportunities for identifying best practices in the different construction methodologies. *Hard costs* are defined as all labor and materials (including overhead and profit) associated with actual construction, and *soft costs* are defined as all labor costs (including overhead and profit) associated with engineering, construction management, and costs of financing the overall project. This study does not examine the differences in program costs or utility tie-in costs because the methods used for these costs were similar on all three projects. The RCC project was completed one month early and below cost, indicating that prescriptive requirements were not needed for successful construction of commercial-type projects. Table 13.1 summarizes the differences in how the projects were executed, and Table 13.2 provides a cost comparison of the three projects based on gross square footage.

Table 13.1 Differences in Execution Methodology Summary

Major ECD Projects	Construction Execution Methodology
RCC Project	Privately developed, commercially funded fast-track, design-build
JICS/ORCAS	State-funded design-bid-build
RSC	DOE-funded line-item design-bid-build project

Table 13.2 Comparison of Cost Data

	RCC		JICS/ORCAS		RSC	
	SF/Cost ($K)	$/ft²	SF/Cost ($K)	$/ft²	SF/Cost ($K)	$/ft²
Gross square footage	376,052 ft²		52,250 ft²		50,140 ft²	
Soft cost	20,613	55	1,611	31	3,138	63
Hard cost[a]	52,712	140	7,300	140	10,346	206
Costs removed to normalize	6,300	17	1,800	34	2,416	48
Total cost	79,625	212	10,711	205	15,900	317

[a] Hard cost = direct cost.

The cost comparisons are based on a combination of actual costs as well as estimates to complete that include fixed price bids and assignment of any remaining contingency. Hard costs do not include special equipment or furniture for the facility.

In this chapter, we present a combined graphical-analytical methodology to compare cost profiles of the construction projects. Cost-effective practices from commercial construction are adopted as baselines against which government projects can be benchmarked. Although the basis for the development of the methodology is an actual study of new government construction projects at ORNL, this chapter focuses only on a presentation of the cost benchmarking methodology itself, rather than on the operational aspects of the projects.

13.3 Basis of the methodology

The study methodology consisted of the following steps:

- Gather technical (scope), cost, and schedule data for each of the three projects by interviewing the responsible project personnel from UT–Battelle, the private development corporation, and construction contractors.
- Identify the differences in execution of the construction projects.
- Develop a cost model for each project, including attributes for benchmarking.

- Benchmark and compare each project's hard and soft costs, identify differences, develop a hypothesis for cost differences, and test the hypothesis to determine if the data and information support the data.
- Develop conclusions and recommendations for improvements.

13.3.1   Data gathering and analysis

The primary sources of data for the analysis were anecdotal, empirical, and analytical.

- **Anecdotal approach:** Based on observation and experiential conjectures that are not necessarily backed by proof, the anecdotal approach is the least reliable.
- **Empirical approach:** Based on an analysis of actual data, this is a verifiable approach backed by observed, recorded data, requiring cooperation and input from constructors.
- **Analytical approach:** Based on building a mathematical or representative quantitative mode, it is derived from all of the cost models and is used for extrapolation as a comparison tool.

To obtain anecdotal data, the research team interviewed project personnel from UT–Battelle, from the private development corporation, and from construction contractors. To obtain empirical data, the research team reviewed baseline defining documents, cost reports, and procurement documents for the subject projects. A survey of published literature on the construction industry was also performed. The anecdotal and empirical data assembled are integrated, presented, and analyzed as appropriate in the sections that follow.

13.3.2   New project financing options

An experiment that has been used successfully at ORNL involves an integrative financing strategy, whereby new buildings at a government site are financed in a tripartite funding partnership of federal government funds, state funds, and commercial financing funds. Table 13.3 presents a summary of the funding partnerships. The commercially developed project provides

Table 13.3 Construction Projects and Funding Sources

Bond (Private) Market Funding
Building Project A
State Capital Funding
Building Project B
Federal Government Capital Project Funding
Building Project C

the best-practice benchmarks for cost and project procedures. The project developed under state funding straddles the typical reputation of government projects and private commercial projects.

13.3.3 Analysis of funding options

The proposed methodology is designed to examine and document the sources of cost differences between the different financing methods for the building projects. The authors carried out a comparative study, which specifically compared Building Project A to Building Project C, using Building Project B as the comparative bridge.

13.3.4 Cost comparison basis

The focus of the analysis was on hard costs only because it was determined that the soft costs contained subtle aspects that were difficult to quantify on a normalized basis. In order to prevent having to compare "apples to oranges," the authors eliminated soft costs from the analysis.

Figure 13.1 shows the building cost relationships. Because Buildings B and C had similar square footage, construction, and material cost, B was used to test the cost of C. Similarly, because B and A both used commercial construction approaches, the cost of B was tested against the cost of A, which was then extrapolated for the required comparison of A to C.

Figure 13.2 shows the typical breakdown of project costs into hard and soft costs. In this particular project, the focus was on hard costs only. Figure 13.3

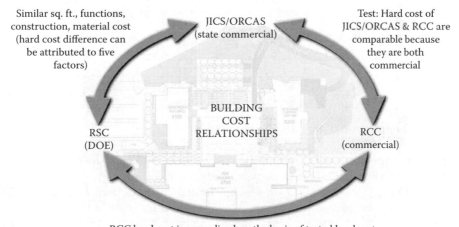

RCC hard cost is normalized on the basis of tested hard cost from the RSC–JICS comparison & JICS/ORCAS–RCC comparison. Therefore, we can use the five factors from RSC-JICS/ORCAS to explain (distribute) the $66 difference over the five factors.

Figure 13.1 Building cost relationships.

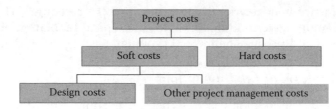

Figure 13.2 Breakdown of project costs.

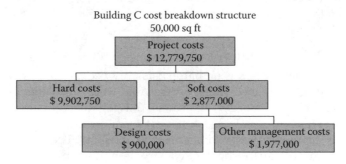

Figure 13.3 Cost breakdown for Building C.

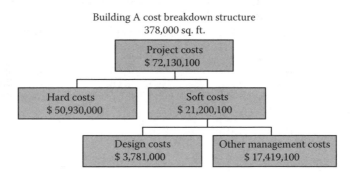

Figure 13.4 Cost breakdown for Building A.

shows the specific breakdown of the cost of Building C into its components. Figure 13.4 shows the breakdown for the cost of Building A.

The hard cost difference to be distributed over the five factors is $63, which is the per-square-foot difference between Building C and Building A. The objective is to identify and analyze the factors responsible for the cost differences.

13.3.5 *Analysis of the cost drivers*

The research team narrowed the cost drivers into five primary factors that could account for the cost differences. The factors are summarized as follows:

Factor 1 (F1): Impact of Davis Bacon wages

Factor 2 (F2): Impact of Project Labor Agreement (PLA)

Factor 3 (F3): Additional environment, safety, health and quality (ESH&Q) requirements, Oversight, Safety Meetings, Site Training

Factor 4 (F4): Excessive requirements of contract terms and conditions

Factor 5 (F5): Impact of procurement process

These factors represent a consolidation of both major and subtle cost drivers in the construction projects.

13.4 Evaluation baselines

To facilitate an effective analysis of the cost comparisons, two baselines were selected. The average of the cost of the commercial development (Building A) facility represents one baseline. The cost of Building C, a federal-funded facility, represents the other baseline. The baselines are summarized in Table 13.4.

For a complete analysis, we suggest a comprehensive benchmarking approach of comparing building construction performance on the basis of the following:

- Hard cost/sq. ft.
- Soft cost/sq. ft.
- Total cost/sq. ft.
- Slippage days/sq. ft.
- Hours of construction/sq. ft.

Graphical representations of the benchmarking comparisons are shown in Figure 13.5, Figure 13.6, and Figure 13.7. For the purpose of this chapter,

Table 13.4 Analysis of Cost Drivers from Building A to Building C

X2	– –	Cost of A	
	Factor 1: Davis Bacon wages		Assess impact of wages
	Factor 2: Project labor agreement		Add impacts of PLA
	Factor 3: Additional ESH&Q		Add the hypothesized impact of additional requirements
	Factor 4: Contract terms and conditions		Add the impact of cost increases due to contract terms and conditions
	Factor 5: Procurement process		Add the cost impact of the government procurement process
X1	– –	Cost of C	Arrive at cost of C

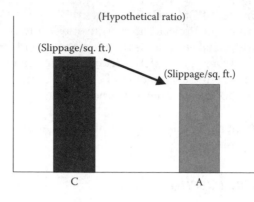

Figure 13.5 Benchmarking measure using slippage/sq. ft.

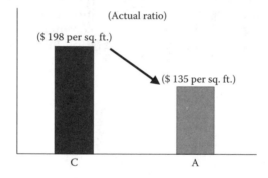

Figure 13.6 Benchmarking measure using cost/sq. ft.

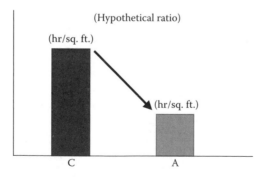

Figure 13.7 Benchmarking measure using hours/sq. ft.

we limit our comparison only to the benchmark measure in Figure 13.6 (hard cost per square foot). With appropriate data, the other measures can also be studied.

Comparative Analysis: The normalized hard cost of Building C is $198 per square foot, whereas the normalized hard cost of A is $135 per square foot. The cost of C is designated as X1, whereas the average cost of A is designated as X2. The proposed methodology establishes the cost drivers between X1 and X2. The general equation relating the baselines can be represented as follows:

X1 = **X2** + Effect of Factor 1 + Effect of Factor 2 + ... + Effect of Factor N

where
$$X1 = \$198 \text{ per square foot}$$
$$X2 = \$135 \text{ per square foot}$$
$$X1 - X2 = \$63$$

This is the amount that needs to be distributed over the cost drivers (factors). The methodology uses a graphical representation approach as a means to implement an empirical approach to allocate the cost difference.

13.5   Data collection processes

Data collection for this study spanned discussions with project stakeholders, contractors, workers, and reference to published literature on the construction industry. The summary of findings reported in this section is based on an integration of the different sources of information. As mentioned earlier, the primary sources of data for the analysis were anecdotal, empirical, and analytical. A part of the empirical data analysis involves a review of the labor loading on the Building C project site. The authors fitted an S-curve to the labor-loading data. The result is shown in Figure 13.8. This agrees with the standard S-curve expectation for project resource loading. With this plot, labor loading can be studied for resource-leveling purposes.

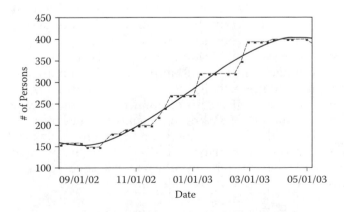

Figure 13.8 S-curve fitted to manpower loading for Building C.

13.6   Analysis of the factors

13.6.1   F1: Impact of Davis Bacon Wages

13.6.1.1   Data findings

- Building C is all union.
- Union wage scales are a little higher than Davis-Bacon wages. In the past, Davis Bacon's wages were higher than commercial wages, but commercial contractors have raised wages to become competitive in retaining good workers.
- 100% of C is subcontracted work, paying prevailing (union) wages.
- The flat nonunion rate is lower. Nonunion labor tends to be more cost-effective for contractors because workers can be given more diversified work assignments.

The fact that C uses all-union labor would suggest the possible existence of cost escalation due to union bargaining. Based on survey inputs of the opinions of construction experts, an average of a 7% impact was assigned to this factor. Thus, we have

$$\text{F1 Impact} = 7\%(\$63) = \$4.41$$

13.6.2   F2: Impact of PLA

PLA affects cost primarily because of jurisdictional work rules for each craft as well as required ratios of journeymen to apprentices.

13.6.2.1   Data findings

According to interviews with contractors, PLA does not seem to affect work output and does not create obvious construction cost differences for the buildings. There are, however, some sources of anecdotal information relating to where and how PLA can adversely affect project cost and productivity. PLA is a stipulation that contractors pay collective prevailing wage scales as a way to promote quality of workmanship, efficiency, and peaceful labor relations. Unfortunately, in some instances, this blanket requirement has led to labor complacency, productivity loss, and cost increases. Previous research has revealed that construction costs could increase and worker productivity could decrease as a result of PLA even if quality remains the same. Previous survey results show that PLA decreases contractor productivity. The prevailing labor referral system and work rules are seen as the primary sources of adverse impacts on productivity. Experiential data suggests an impact range of around 47%. Thus, we have the following:

$$\text{F2 Impact} = 47\%(\$63) = \$29.61$$

13.6.3   F3: Additional ESH&Q requirements

There are additional ESH&Q requirements that affect new building projects at ORNL. These additional oversight requirements often create additional costs. It is generally agreed that differences of scope among different buildings generate differences in cost per square foot. This can be reflected quantitatively in RS Means calculations and other data. The designed use of a building may dictate the type of oversight, safety requirements, materials, wall thickness, floor layout, carpeting, ceilings, conference room facilities, and so on, with the consequence of cost increases.

13.6.3.1   Data findings

The project team used RS Means calculations for this factor. Building size is one factor that affects the square-foot cost. The larger facility will have a lower square-foot cost due to the decreasing cost contribution of the exterior walls plus the economy of scale achievable in larger buildings.

13.6.3.1.1   RS means calculations.   RS Means were computed for the buildings on a comparative basis. Area conversion factors from the RS Means Handbook were used to convert costs for a typical-size building to an adjusted cost for a particular building project. For the privately developed buildings we have the following:

Total square footage = 376,000 sq. ft.
Assume the following square-footage allocations:
    A = 1/3 research space in Building A = 125,333 sq. ft.
    B = 1/3 computer space in Building A = 125,333 sq. ft.
    C = 1/3 office space in Building A = 125,333 sq. ft.

A: Research Space
Means Median = $139.00/sq. ft.
Typical sq. ft. = 19,000
Therefore, proposed sq. ft./typical sq. ft. = 125,333/19,000 = 6.6 > 3.
Therefore, we will use a factor of 0.90.
Thus, 125,333 ∞ 139 ∞ 0.90 = 15,679,158.

B: Computer Space: 15,679,158

C: Office Space:
Means Median = $79.45 sq. ft.
Typical sq. ft. = 20,000.
Therefore, proposed sq. ft./typical sq. ft = 125,333/20,000 = 6.15 > 3.
Therefore, use 0.90 factor.
Thus, 125,333 ∞ 79.45 ∞ 0.90 = 8,961,936.
A + B + C = 40,320,252.

A+B+C/376,000 = 40,320,252/376,000 = \$107.23/sq. ft., which is the expected estimated average cost per square foot for the commercial buildings.

For Building C:
Assume A = 16,000 sq. ft. for restaurant.
B = 37,000 sq. ft. for college student union.

A: Means Median = \$119/sq. ft.
Typical sq. ft. = 4,400
Proposed sq. ft/typical sq. ft. = 16,000/4,400 = 3.63 > 3.
Therefore, use 0.90 factor.
16,000 sq. ft. ∞ 119 ∞ 0.90 = 1,713,600.

B: Means Median = \$129/sq. ft.
Typical sq. ft. = 33,400.
Proposed sq. ft./typical sq. ft. = 37,000/33,400 = 1.1 < 3.
Therefore, use 0.98 factor.
Thus, 37,000 ∞ 129 ∞ 0.98 = 4,773,000.
A+B = 6,486,600.
A+B/53,000 = 6,486,600/53,000 = 122.4.

Now, (122.40 – 107.23)/107.23 = 14.15%, which implies a 14.15% cost increase of C over Building A. Thus, we have

$$F3 \text{ Impact} = 14\%(\$63) = \$8.82$$

where 14.15% has been rounded down to 14%.

13.6.4   F4: Impact of excessive contract terms and conditions

Government contract terms and conditions often have excessive requirements that have cost implications. Based on experiential data obtained from construction supervisors, we allocate a 14% impact to this factor. Thus, we have

$$F4 \text{ Impact} = 14.00\%(\$63) = \$8.82$$

13.6.5   F5: Impact of procurement process

One nuance of government-funded projects is the cumbersome procurement process, which leads to less cost-effective implementations. A suggested impact of 18% was assigned to this factor. Thus, we have

$$F5 \text{ Impact} = 18.00\%(\$63) = \$11.34$$

Table 13.5 Summary of Cost Difference Distribution

Factor	Percentage (%)	Dollar Amount
1	7.00	4.41
2	47.00	29.61
3	14.00	8.82
4	14.00	8.82
5	18.00	11.34
Total	**100.00**	**$63.00**

A summary of the cost distribution over the factors is shown in Table 13.5. A graphical representation of the cost difference contribution is shown in Figure 13.9. Figure 13.10 presents the graphical information in a bar-chart format.

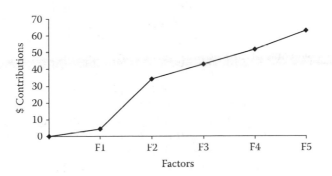

Figure 13.9 Line chart of factor contributions by Dollar amounts.

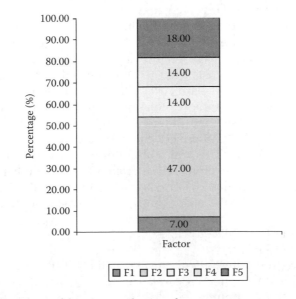

Figure 13.10 Bar chart of factor contributions by percentage.

13.7   Conclusions and recommendations

The results in this study are validated based on a combination of anecdotal information, quantitative computations, a review of construction data, and reference to published literature on government construction projects.

The conclusion is that government construction projects, by virtue of their bureaucratic nature, create an incidence of cost escalations and productivity loss. It is recommended that government projects be benchmarked against commercial projects. In this respect, government projects can borrow the "best practices" of commercially executed projects in lowering construction costs and improving work productivity.

Most contractors will carry out their projects successfully if given operating guidelines and left with the latitude to develop the means and methods to reduce cost and improve quality and productivity. In spite of the prevailing tight government control procedures, management oversight, and implementation audits, many government projects still end up with quality problems and cost overruns. Project analysts should review the marginal benefit of government procedures vs. the incremental cost brought on by those procedures. The government process itself has fueled the stereotype that government projects have infinite loopholes and unlimited funding potentials. This has the consequence that contractors, who would have normally executed projects under their own efficient methods, resort to unproductive practices in order to take advantage of the "government project" stereotypes and extra opportunities for profit.

Due to the difference in project cost accounting and cost collection methods used on these projects, the soft costs could not be compared in a meaningful way. However, design costs as a percentage of construction costs were compared for the projects involved, and they all fall into a range of 6 to 7%. Although the reasons for the differences in the soft costs were not evaluated in this study, anecdotal evidence suggests they may be caused by overly prescriptive requirements and lengthy design review cycles required for DOE projects. These soft costs need to be evaluated in a separate focused study. Table 13.2 depicts the hard costs of the RSC (DOE-funded project) as approximately $66 per sq. ft higher than the other two construction projects.

To further understand the differences in hard costs, the RSC project (DOE) and the JICS/ORCAS project (State of Tennessee/commercial) were compared in more detail. This was straightforward because of their similar square footage, similar structures, close proximity to each other, and similar contracting (design–bid–build). In comparing the two, the following five factors were found to be the major contributors to the hard cost differences:

- Impact of Davis-Bacon wages
- Impact of PLA
- Overly prescriptive requirements of the UT–Battelle Construction Specification Division 1, *General Requirements*, including ESH&Q oversight, and site training

- Apparently excessive requirements of procurement contract terms and conditions
- Impact of lengthy procurement process on project schedules

The available data confirm that the DOE approach to new commercial or light industrial (nonnuclear) buildings is process driven. This emphasis forces contractors to focus on adhering to the requirements of DOE documents (*Processes for Project Management* [Required by DOE O 413.3] and *Construction Project Acquisition*), rather than focusing on the successful construction of a building. Thus, a project is considered more successful if the DOE process has been followed without any mistakes, rather than focusing on assuring that the maximum building (bricks and mortar) is constructed for the allocated budget. The contractor is concerned more about satisfying the process instead of pursuing practices that will achieve building performance that meets or exceeds expectations.

13.7.1   Recommendations for reducing future project costs

- Perform further benchmarking to determine commercial practices and implement changes as appropriate to further reduce construction costs.
- Recommend to DOE the elimination of prescriptive requirements that are not required by regulations and statutes, and the identification of ways to change facility construction progress from a process-based project management system to a performance-based project management system.
- Evaluate the need for a PLA for building federal and/or DOE "commercial-like" construction projects.
- Evaluate the need for a standard DOE line-item project management process and review cycle according to DOE O 413.3, *Program and Project Management for the Acquisition of Capital Assets*, for commercial-type projects under $50 million. The benefit of using the standard project management process for research or science projects is questionable.
- Recommend that DOE use a more commercial, less process-driven approach for new construction projects (e.g., eliminate DOE O 413.3 for "nonnuclear commercial-like" projects). For example, the research team recommends fewer reviews, focusing on performance, and focusing on the use of commercial contracting methods (e.g., American Institute of Architects [AIA] model contracts). These approaches result in more facility for the dollar and better "other" performance indicators (e.g., safety performance). The research team also recommends the use of DOE project personnel to support and provide oversight of the management and operating (M&O) contractor in adopting this revised approach.
- The team recommends further benchmarking studies at other locations to confirm the data and to provide the DOE with additional rationale for adapting its processes for commercial-type construction projects.

13.7.1.1 Overall project recommendation

For regular construction not involving safety-critical process plants, best commercial construction practices should be used. The practice of using the same rigid review process for all projects is wasteful and ineffective. Non-process-plant projects do not need tight review and control of every project step. It is recommended that other government construction projects be cost-benchmarked by using a methodology similar to the one presented in this chapter.

Reference

1. Badiru, Adedeji B., Delgado, Vincent, Ehresma-Gunter, Jamie, Omitaomu, Olufemi A., Nsofor, Godswill, and Saripali, Sirisha (2004). Graphical and Analytical Methodology for Cost Benchmarking for New Construction Projects. *Proceedings of 13th International Conference on Management of Technology (IAMOT 2004)*, Washington, DC. April 3–7.

appendix A

Definitions and terms

Accounting Life: The period of time over which the amount of the asset cost to be depreciated, or recovered, will be allocated to expenses by accountants.

Actual Dollars: Cash flow at the time of the transaction.

Alternative, Contingent: An alternative that is feasible only if some other alternative is accepted. The opposite of a mutually exclusive alternative.

Alternative, Economic: A plan, project, or course of action intended to accomplish some objective that has or will be valued in monetary terms.

Alternative, Independent: An alternative such that its acceptance has no influence on the acceptance of other alternatives under consideration.

Alternative, Mutually Exclusive: An alternative such that its selection rules out the selection of any other alternatives under consideration.

Amortization: (1) (a) As applied to a capitalized asset, the distribution of the initial cost by periodic charges to expenses as in depreciation. Most amortizable assets have no fixed life; (b) The reduction of a debt by either periodic or irregular payments. (2) A plan to pay off a financial obligation according to some prearranged program.

Annual Cost: The negative of Annual Worth. (See Equivalent Uniform Annual Cost.)

Annual Equivalent: In the time value of money, one of a sequence of equal end-of-year payments that would have the same financial effect when interest is considered as another payment or sequence of payments that are not necessarily equal in amount or equally spaced in time.

Annual Worth: A uniform amount of money at the end of each and every period over the planning horizon, equivalent to all cash flows discounted over the planning horizon when interest is considered.

Annuity: (1) An amount of money payable to a beneficiary at regular intervals for a prescribed period of time out of a fund reserved for that purpose. (2) A series of equal payments occurring at equally spaced periods of time.

221

Annuity Factor: The function of interest rate and time that determines the amount of periodic annuity that may be paid out of a given fund. (See Capital Recovery Factor.)

Annuity Fund: A fund that is reserved for payment of annuities. The present worth of funds required to support future annuity payments.

Annuity Fund Factor: The function of the interest rate and time that determines the present worth of funds required to support a specified schedule of annuity payments. (See Present Worth Factors, Uniform Series.)

Apportion: In accounting or budgeting, the process by which a cash receipt or disbursement is divided among and assigned to specific time periods, individuals, organization units, products, projects, services, or orders.

Bayesian Statistics: (1) Classical — the use of probabilistic prior information and evidence about a process to predict probabilities of future events. (2) Subjective — the use of subjective forecasts to predict probabilities of future events.

Benefit/Cost (Cost-Benefit) Analysis: An analysis technique in which the consequences on an investment evaluated in monetary terms are divided into separate categories of costs and benefits. Each category is then converted into an annual equivalent or present worth for analysis purposes.

Benefit/Cost Ratio: A measure of project worth in which the equivalent benefits are divided by the equivalent costs.

Benefit/Cost Ratio Method: (See Benefit-Cost Analysis.)

Book Value: The original cost of an asset or group of assets less the accumulated book depreciation.

Break-Even Chart: (1) A graphic representation of the relation between total income and total costs (sum of fixed and variable costs) for various levels of production and sales indicating areas of profit and loss. (2) Graphic representation of a figure of merit as a function of a specified relevant parameter.

Break-Even Point: (1) The rates of operations, output, or sales at which income will just cover costs. Discounting may or may not be used in making these calculations. (2) The value of a parameter such that two courses of action result in an equal value for the figure of merit.

Capacity Factor: (1) The ratio of current output to maximum capacity of the production unit. (2) In electric utility operations, it is the ratio of the average load carried during a period of time divided by the installed rating of the equipment carrying the load. (See Demand Factor and Load Factor).

Capital: (1) The financial resources involved in establishing and sustaining an enterprise or project. (2) A term describing wealth that may be utilized to economic advantage. The form that this wealth takes may be as cash, land, equipment, patents, raw materials, finished products, etc. (See Investment and Working Capital.)

Capital Budgeting: The process by which organizations periodically allocate investment funds to proposed plans, programs, or projects.

Capital Recovery: (1) Charging periodically to operations amounts that will ultimately equal the amount of capital expended. (2) The replacement of the original cost of an asset plus interest. (3) The process of regaining the new investment in a project by means of setting revenues in excess of the economic investment costs. (See Amortization, Depletion, and Depreciation.)

Capital Recovery Factor: A number that is a function of time and the interest rate, used to convert a present sum to an equivalent uniform annual series of end-of-period cash flows. (See Annuity Factor.)

Capital Recovery with Return: The recovery of an original investment with interest. In the public utility industry, this is frequently referred to as the revenue requirements approach.

Capitalized Asset: Any asset capitalized on the books of account of an enterprise.

Capitalized Cost: (1) The present worth of a uniform series of periodic costs that continue indefinitely (hypothetically infinite). Not to be confused with capitalized expenditure. (2) The present sum of capital which, if invested in a fund earning a stipulated interest rate, will be sufficient to provide for all payments required to replace and/or maintain an asset in perpetual service.

Cash Flow: The actual monetary units (e.g., dollars) passing into and out of a financial venture or project being analyzed.

Cash-Flow Diagram: The illustration of cash flows (usually vertical arrows) on a horizontal line where the scale along the line is divided into time-period units.

Cash-Flow Table: A listing of cash flows, positive and negative, in a table in the order of the time period in which the cash flow occurs.

Challenger: In replacement analysis, a proposed property or equipment that is being considered as a replacement for the presently owned property or equipment (the defender). In the analysis of multiple alternatives, an alternative under consideration that is to be compared with the last acceptable alternative (the defender). (See MAPI Method.)

Common Costs: In accounting, costs that cannot be identified with a given output of products, operations, or services. Expenditures that are common to all alternatives.

Compound Amount: (1) The equivalent value, including interest, at some stipulated time in the future of a series of cash flows occurring prior to that time. (2) The monetary sum that is equivalent to a single (or a series of) prior sums when interest is compounded at a given rate.

Compound Amount Factors: Functions of interest and time that when multiplied by a single cash flow (single payment compound amount factor), or a uniform series of cash flows (uniform series compound amount factor) will give the future worth at compound interest of such single cash flow or series.

Compound Interest: (1) The type of interest that is periodically added to the amount of investment (or loan) so that subsequent interest is based on the cumulative amount. (2) The interest charges under the condition that interest is charged on any previous interest earned in any time period, as well as on the principal.

Compounding, Continuous: A compound-interest assumption in which the compounding period is of infinitesimal length and the number of periods is infinitely great. A mathematical concept that is conceptually attractive and mathematically convenient for dealing with frequent (e.g., daily) compounding periods within a year.

Compounding, Discrete: A compound interest assumption in which the compounding period is of a specified length such as a day, week, month, quarter year, half year, or year.

Compounding Period: The time interval between dates (or discrete times) at which interest is paid and added to the amount of an investment or loan. Usually designates the frequency of compounding during a year.

Constant Dollars: An amount of money at some point in time, usually the beginning of the planning horizon, equivalent in purchasing power to the active dollars necessary to buy the good or service. Actual dollars adjusted to a relative price change.

Cost/Benefit Analysis: (See Benefit/Cost Analysis.)

Cost-Effectiveness Analysis: An analysis in which the major benefits may not be expressed in monetary terms. One or more effectiveness measures are substituted for monetary values, resulting in a trade-off between marginal increases in effectiveness and marginal increases in costs.

Cost of Capital: A term, usually used in capital budgeting, to express as an interest rate percentage the overall estimated cost of investment capital at a given point in time, including both equity and borrowed funds.

Current Dollars: (See Actual Dollars).

Cutoff Rate of Return (Hurdle Rate): The rate of return after taxes that will be used as a criterion for approving projects or investments. It is determined by management based on the supply and demand for funds. It may or may not be equal to the minimum attractive rate of return (MARR) but is at least equal to the estimated cost of capital.

Decision Theory: With reference to engineering economy, it is a branch of economic analysis devoted to the study of decision processes involving multiple possible outcomes, defined either discretely or on a continuum, and deriving from the theory of games and economic behavior and probabilistic modeling.

Decision Tree: In decision analysis, a graphical representation of the anatomy of decisions showing the interplay between a present decision, the probability of chance events, possible outcomes and future decisions, and their results or payoffs.

Decisions Under Certainty: From the literature of decision theory, that class of problems wherein single estimates with respect to cash flows and economic life (complete information) are used in arriving at a decision among alternatives.

Decisions Under Risk: From the literature of decision theory, that class of problems in which multiple outcomes are considered explicitly for each alternative, and the probabilities of the outcomes are assumed to be known.

Decisions Under Uncertainty: From the literature of decision theory, that class of problems in which multiple outcomes are considered explicitly for each alternative but the probabilities of the outcomes are assumed to be unknown.

Defender: In replacement analysis, the presently owned property or equipment being considered for replacement by the most economical challenger. In the analysis of multiple alternatives, the previously judged acceptable alternative against which the next alternative to be evaluated (the challenger) is to be compared.

Deflating (By a Price Index): Adjusting some nominal magnitude, e.g., an actual dollar estimate, by a price index. It may be required in order to express that magnitude in units of constant purchasing power. (See Inflating, Constant Dollars, and Current Dollars.)

Deflation: A decrease in the relative price level of a factor of production, an output, or the general price level of all goods and services. A deflationary period is one in which there is (or is expected to be) a sustained decrease in price levels.

Demand Factor: (1) The ratio of the current production rate of the system divided by the maximum instantaneous production rate. (2) The ratio of the average production rate, as determined over a specified period of time, divided by the maximum production rate. (3) In electric utility operations, it is the ratio of the maximum kilowatt load demanded during a given period divided by the connected load. (Also see Capacity Factor and Load Factor.)

Depletion: (1) An estimate of the lessening of the value of an asset due to a decrease in the quantity available for exploitation. It is similar to depreciation except that it refers to natural resources such as coal, oil, and timber. (2) A form of capital recovery applicable to properties such as listed previously. Its determination for income-tax purposes may be on a unit of production basis, related to original cost or appraised value of the resource (known as cost depletion), or based on a percentage of the income received from extracting or harvesting (known as percentage depletion.)

Depletion Allowance: An annual tax deduction based upon resource extraction. (See Depletion.)

Depreciation: (1) (a) Decline in value of a capitalized asset; (b) A form of capital recovery, usually without interest, applicable to property with two or more years' life span in which an appropriate portion of the asset's value periodically is charged to current operations. (2) A loss of value due to physical or economic reasons. (3) In accounting, depreciation is the allocation of the book value of this loss to current operations according to some systematic plan. Depending on then existing income-tax laws, the amount and timing of the charge to current operations for tax purposes may differ from that used to report annual profit and loss.

Depreciation, Accelerated: Depreciation methods accepted by the taxing authority that write off the value (cost) of an asset usually over a shorter period of time (i.e., at a faster rate) than the expected economic life of the asset. An example is the Accelerated Cost Recovery System (ACRS) introduced in the U.S. in 1981 and modified in later years.

Depreciation Allowance: An annual income-tax deduction, and/or charge to current operations, of the original cost of a fixed asset. The income-tax deduction may not equal the charge to current operations. (See Depreciation.)

Depreciation Basis: In tax accounting, the cost of the otherwise-determined value of a group of fixed assets, including installation costs and certain other expenditures, and excluding certain allowances. The depreciation basis is the amount that, by law, and/or acceptance by the taxing authority, may be written off for tax purposes over a period of years.

Depreciation, Declining Balance: A method of computing depreciation in which the annual charge is a fixed percentage of the depreciated book value at the beginning of the year to which the depreciation charge applies.

Depreciation, Multiple Straight-Line: A method of depreciation accounting in which two or more straight-line rates are used. This method permits a predetermined portion of the asset to be written off in a fixed number of years. One common practice is to employ a straight-line rate that will write off Ω of the cost in the first half of the anticipated service life with a second straight-line rate used to write off the remaining π in the remaining half life. (See Depreciation, Straight-Line.)

Depreciation, Sinking Fund: (1) A method of computing depreciation in which the periodic charge is assumed to be deposited in a sinking fund that earns interest at a specified rate. The sinking fund may be real but usually is hypothetical. (2) A method of depreciation where a fixed sum of money regularly is deposited at compound interest in a real or hypothetical fund in order to accumulate an amount equal to the total depreciation of an asset at the end of the asset's estimated life. The depreciation charge to operations for each period equals the sinking fund deposit amount plus interest on the beginning of the period sinking fund balance.

Depreciation, Straight-Line: A method of computing depreciation wherein the amount charged to current operations is spread uniformly over the estimated life of an asset. The allocation may be performed on a unit-of-time basis or a unit-of-production basis, or some combination of the two.

Depreciation, Sum-of-Years Digits: A method of computing depreciation wherein the amount charged to current operations for any year is based on the ratio: (years of remaining life)/(1 + 2 + 3 + ... + n), n being the estimated life.

Deterioration: A reduction in value of a fixed asset due to wear and tear and action of the elements. It is a term used frequently in replacement analysis.

Development Cost: (1) The sum of all the costs incurred by an inventor or sponsor of a project up to the time the project is accepted by those who will promote it. (2) In international literature, activities in developing nations intended to improve their infrastructure.

Direct Cost: A traceable cost that can be segregated and charged against specific products, operations, or services.

Discount Rate: (See Interest Rate and Discounted Cash Flow.)

Discounted Cash Flow: (1) Any method of handling cash flows over time, either receipts or disbursements, in which compound interest and compound interest formulae are employed in their analytical treatment. (2) An investment analysis that compares the present worth of projected receipts and disbursements occurring at designated times in order to estimate the rate of return from the investment or project. In this sense, also see Rate of Return and Profitability Index.

Dollars, Constant (Real Dollars): Dollars (or some other monetary unit) of constant purchasing power independent of the passage of time. In situations where inflationary or deflationary effects have been assumed when cash flows were estimated, those estimates are converted to constant dollars (base-year dollars) by adjustment by some readily accepted general inflation/deflation index. Sometimes termed actual dollars, although this term also is used to describe current dollar values. (See Current Dollars, Inflating, and Deflating.)

Dollars, Current (Then-Current Dollars): Estimates of future cash flows that include any anticipated changes in amount due to inflationary or deflationary effects. Usually, these amounts are determined by applying an index to base-year dollar (or other monetary unit) estimates. Sometimes termed actual dollars, although this term also is used to describe constant dollar values. (See Constant Dollars, Inflating, and Deflating.)

Earning Value (Earning Power of Money): The present worth of an income producer's estimated future net earnings as predicted on the basis of recent and present expenses and earnings and the business outlook.

Economic Life: The period of time extending from the date of installation to the date of retirement from the intended service, over which a prudent owner expects to retain an equipment or property so as to minimize cost or maximize net return. (See Life.)

Economy: (1) The cost or net return situation regarding a practical enterprise or project, as in economy study, engineering economy, or project economy. (2) A system for the management of resources. (3) The avoidance of (or freedom from) waste in the management of resources.

Effective Interest: (See Interest Rate, Effective.)

Effectiveness: In the engineering economy, the measurable consequences of an investment not reduced to monetary terms; e.g., reliability, maintainability, safety.

Endowment: A fund established for the support of some project for succession of donations or financial obligations.

Endowment Method: As applied to an economy study, a comparison of alternatives based on the present worth or capitalized cost of the anticipated financial events.

Engineering Economy: (1) The application of economic or mathematical analysis and synthesis to engineering decisions. (2) A body of knowledge and techniques concerned with the evaluation of the worth of commodities and services relative to their costs and with methods of estimating inputs.

Equivalent Uniform Annual Cost (EUAC): (See Annual Cost.)

Estimate: A magnitude determined as closely as it can be by the use of past history and the exercise of sound judgment based upon approximate computations, not to be confused with offhand approximations that are little better than outright guesses.

Exchange Rate: The rate at a given point in time at which the currency of one nation exchanges for that of another.

Expected Yield: In finance, the ratio of the expected return from an investment, usually on an after-tax basis, divided by the investment.

External Rate of Return: A rate of return calculation that takes into account the cash receipts and disbursements of a project and assumes that all net receipts (cash throw offs) are reinvested elsewhere in the enterprise at some stipulated interest rate. (Also see Rate of Return and Internal Rate of Return.)

Fair Rate of Return: The maximum rate of return that an investor-owned public utility is entitled to earn on its rate base in order to pay interest and dividends and attract new capital. The rate, or percentage, usually is determined by state or federal regulatory bodies.

First Cost: The initial investment in a project or the initial cost of capitalized property, including transportation, installation, preparation for service, and other related initial expenditures.

Fixed Assets: The tangible portion of an investment in an enterprise or project that is comprised of land, buildings, furniture, fixtures, and equipment with an expected life greater than 1 year.

Fixed Cost: Those costs that tend to be unaffected by changes in the number of units produced or the volume of service given.

Future Worth: (1) The equivalent value at a designated future date based on the time value of money. (2) The monetary sum, at a given future time, which is equivalent to one or more sums at given earlier times when interest is compounded at a given rate.

Going-Concern Value: The difference between the value of a property as it stands possessed of its going elements and the value of the property alone as it would stand at completion of construction as a bare or inert assembly of physical parts.

Goodwill Value: That element of value that inheres in the fixed and favorable consideration of customers arising from an established well-known and well-conducted business. This is determined as the difference between what a prudent business person is willing to pay for the property and its going-concern value.

Gradient Factors: A group of compound-interest factors used for equivalence conversions of arithmetic or geometric gradients in cash flow. In general use are the arithmetic gradient to uniform series (gradient conversion) factor, the arithmetic gradient to present worth (gradient present worth) factor, and the geometric gradient to present worth factor.

Increment Cost (Incremental Cost): The additional (or direct) cost that will be incurred as the result of increasing output by one unit more. Conversely, it may be defined as the cost that will not be incurred if the output is reduced by one unit. (2) The variation in output resulting from a unit change in input. (3) The difference in costs between a pair of mutually exclusive alternatives.

Indirect Cost: Nontraceable or common costs that are not charged against specific products, operations, or services but rather are allocated against all (or some group of) products, operations, and/or services by a predetermined formula.

Inflating (By a Price Index): The adjustment of a present- or base-year price by a price index in order to obtain an estimate of the current (or then current) price at future points in time. (See Deflating, Constant Dollars, and Current Dollars.)

Inflation: A persistent rise in price levels, generally not justified by increased productivity, and usually resulting in a decline in purchasing power. Sometimes the term is used interchangeably with escalation. However, this latter term more often is restricted to the differential increase in a price relative to some specific index of general changes in price levels. (See Deflation.)

Intangibles: (1) In economy studies, those elements, conditions, or economic factors that cannot be evaluated readily or accurately in monetary terms. (2) In accounting, the assets of an enterprise that cannot reliably be values in monetary terms (e.g., goodwill). (See Irreducibles.)

Interest: (1) The monetary return or other expectation that is necessary to divert money away from consumption and into long-term investment. (2) The cost of the use of capital. It is synonymous with the term time value of money. (3) In accounting and finance, (a) a financial share in a project or enterprise; (b) periodic compensation for the lending of money.

Interest Rate: The ratio of the interest accrued in a given period of time to the amount owed or invested at the start of that period.

Interest Rate, Effective: The actual interest rate for one specified period of time. Frequently, the term is used to differentiate between nominal annual interest rates and actual annual interest rates when there is more than one compounding period in a year.

Interest Rate, Market: The rate of interest quoted in the market place that includes the combined effects of the earning value of capital, the availability of funds, and anticipated inflation or deflation.

Interest Rate, Nominal: (1) The interest rate for some period of time that ignores the compounding effect of interest calculations during sub-periods within that period. (2) The annual interest rate, or annual percentage rate (APR), frequently quoted in the media.

Interest Rate, Real: An estimate of the true earning rate of money when other factors, especially inflation, affecting the market rate have been removed.

Internal Rate of Return: A rate of return calculation which takes into account only the cash receipts and disbursements generated by an investment, their timing, and the time value of money. (See Rate of Return, Discounted Cash Flow, and External Rate of Return.)

Investment: (1) As applied to an enterprise as a whole, the cost (or present value) of all the properties and funds necessary to establish and maintain the enterprise as a going concern. The capital tied up in an enterprise or project. (2) Any expenditure which has substantial and enduring value (generally more than one year) and which is therefore capitalized. (See First Cost.)

Investor's Method: A term most often used in the valuation of bonds. (See Rate of Return, Internal Rate of Return, and Discounted Cash Flow.)

Irreducibles: Those intangible conditions or economic factors which cannot readily be reduced to monetary terms; e.g., ethical considerations, esthetic values, or nonquantifiable potential environment concerns.

Leaseback: A business arrangement wherein the owner of land, buildings, and/or equipment sells such assets and simultaneously leases them back under a long-term lease.

Life: (1) Economic: that period of time after which a machine or facility should be retired from primary service and/or replaced as determined by an engineering economy study. The economic impairment may be absolute or relative. (2) Physical: that period of time after which a machine or facility can no longer be repaired or refurbished to a level such that it can perform a useful function. (3) Service: that period of time after which a machine or facility cannot perform satisfactorily its intended function without major overhaul.

Life Cycle Cost: The present worth or equivalent uniform annual cost of equipment or a project which takes into account all associated cash flows throughout its life including the cost of removal and disposal.

Load Factor: (1) Applied to a physical plant or equipment, it is the ratio of the average production rate for some period of time to the maximum rate. Frequently, it is expressed as a percentage. (2) In electric utility operations, it is the average electric usage for some period of time divided by the maximum possible usage. (See Capacity Factor and Demand Factor.)

MAPI Method: A procedure for equipment replacement analysis developed by George Terborgh for the Machinery and Allied Products Institute. It uses a fixed format and provides charts and graphs to facilitate calculations. A prominent feature of this method is that it includes explicitly an allowance for obsolescence.

Marginal Cost: (1) The rate of change of cost as a function of production or output. (2) The cost of one additional unit of production, activity, or service. (See Increment Cost and Direct Cost.)

Matheson Formula: A title for the formula used for declining balance depreciation. (See Declining Balance Depreciation.)

Maximax Criterion: In decision theory, probabilities unknown, is a rule that says choose the alternative with the maximum of the maximum returns identified for each alternative.

Maximin Criterion: In decision theory, probabilities unknown, is a rule that says choose the alternative with the maximum of the minimum returns identified for each alternative. Also called a maximum security level strategy or Wald's strategy.

Minimax Criterion: In decision theory, probabilities unknown, is a rule that says choose the alternative with the minimum of the maximum costs identified for each alternative. Also called a maximum security level strategy.

Minimax Regret Criterion: In decision making under uncertainty, a rule that says choose the alternative with the least potential net return or cost regret.

Minimin Criterion: In decision theory, probabilities unknown, is a rule that says choose the alternative with the minimum of the minimum costs identified for each alternative.

Minimum Attractive Rate of Return: The effective annual rate of return on investment, either before or after taxes, which just meets the investor's threshold of acceptability. It takes into account the availability and demand for funds as well as the cost of capital. Sometimes termed the minimum acceptable return. (See Cost of Capital and Cutoff Rate of Return.)

Minimum Cost Life: (See Economic Life.)

Multiple Rates of Return (Multiple Roots): A situation in which the structure of a cash-flow time series is such that it contains more than one solving internal rate of return.

Nominal Dollars: (See Actual Dollars.)

Nominal Interest: (See Interest Rate, Nominal.)

Obsolescence: (1) The condition of being out of date. A loss of value occasioned by new developments that place the older property at a competitive disadvantage. A factor in depreciation. (2) A decrease in the value of an asset brought about by the development of new and more economical methods, processes, and/or machinery. (3) The loss of usefulness or worth of a product or facility as the result of the appearance of better and/or more economical products, methods, or facilities.

Opportunity Cost: The cost of not being able to use monetary funds otherwise due to that limited resource being applied to an "approved" investment alternative and thus not being available for investment in other income-producing alternatives. Sometimes expressed as a rate.

Payback Period: (1) Regarding an investment, the number of years (or months) required for the related profit or savings in operating cost to equal the amount of said investment. (2) The period of time at which a machine, facility, or other investment has produced sufficient net revenue to recover its investment costs.

Payback Period, Discounted: Same as Payback Period except the period includes a return on investment at the interest rate used in the discounting.

Payoff Period: (See Payback Period.)

Payoff Table: A tabular presentation of the payoff results of complex decision questions involving many alternatives, events, and possible future states.

Payout Period: (See Payback Period.)

Perpetual Endowment: An endowment with hypothetically infinite life. (See Capitalized Cost and Endowment.)

Planning Horizon: (1) A stipulated period of time over which proposed projects are to be evaluated. (2) That point of time in the future at which subsequent courses of action are independent of decisions made prior to that time. (3) In utility theory, the largest single dollar amount that a decision maker would recommend to be spent. (See Utility.)

Present Worth (Present Value): (1) The monetary sum that is equivalent to a future sum or sums when interest is compounded at a given rate. (2) The discounted value of future sums.

Present Worth Factors: (1) Mathematical formulae involving compound interest used to calculate present worths of various cash-flow streams. In table form, these formulae may include factors to calculate the present worth of a single payment, of a uniform annual series, of an arithmetic gradient, and of a geometric gradient. (2) A mathematical expression also known as the present value of an annuity of one. (The present worth factor, uniform series, also is known as the Annuity Fund Factor.).

Principal: Property or capital, as opposed to interest or income.

Profitability Index: An economic measure of project performance. There are a number of such indexes described in the literature. One of the most widely quoted is one originally developed and so named (the PI) by Ray I. Reul, which essentially is based upon the internal rate of return. (Also see Discounted Cash Flow, Investor's Method and Rate of Return).

Promotion Cost: The sum of all expenses found to be necessary to arrange for the financing and organizing of the business unit that will build and operate a project.

Rate of Return (Internal Rate of Return): (1) The interest rate earned by an investment. (2) The interest rate at which the present worth equation (or the equivalent annual worth or future worth equations) for the cash flows of a project or project increment equals zero. (3) As used in accounting, often it is the ratio of annual profit, or average annual profit, to the initial investment or the average book value.

Rate of Return, External: A rate of return calculation that employs one or more supplemental interest rates to produce equivalence transformations on a portion or all of the cash flows and then solves for rate of return on that equivalent cash-flow series.

Real Dollars: (See Constant Dollars.)

Replacement Policy: A set of decision rules for the replacement of facilities that wear out, deteriorate, become obsolete, or fail over a period of time. Replacement models generally are concerned with comparing the increasing operating costs (and possibly decreasing revenues) associated with aging equipment against the net proceeds from alternative equipment.

Replacement Study: An economic analysis involving the comparison of an existing facility and one or more facilities with equal or improved characteristics proposed to supplant or displace the existing facility.

Required Return: The minimum return or profit necessary to justify an investment. Often it is termed interest, expected return or profit, or charge for the use of capital. It is the minimum acceptable percentage, no more and no less. (See Cost of Capital, Cutoff Rate of Return, and Minimum Attractive Rate of Return.)

Retirement of Debt: The termination of a debt obligation by appropriate settlement with the lender. The repayment is understood to be in the full amount unless partial settlement is specified.

Risk: (1) Exposure to a chance of loss or injury. (2) Exposure to undesired economic consequences.

Risk Analysis: Any analysis performed to assess economic risk. Often this term is associated with the use of decision trees. (See Decision Under Risk and Decision Tree.)

Salvage Value: (1) The cost recovered or that could be recovered from a used property when removed from service, sold, or scrapped. A factor used in appraisal of property value and in computing depreciation. (2) Normally, it is an estimate of an asset's net market value at the end of its estimated life. In some cases, the cost of removal may exceed any sale or scrap value; thus, net salvage value is negative. (3) The market value that a machine or facility has at any point in time.

Sensitivity: The relative magnitude of decision criterion change with changes in one or more elements of an economy study. If the relative magnitude of the criterion exhibits large change, the criterion is said to be sensitive; otherwise it is insensitive.

Sensitivity Analysis: A study in which the elements of an engineering economy study are changed in order to test for sensitivity of the decision criterion. Typically, it is used to assess needed measurement or estimation precision, and often it is used as a substitute for more formal or sophisticated methods such as risk analysis.

Service Life: (See Life.)

Simple Interest: (1) Interest that is not compounded, i.e., is not added to the income-producing investment or loan. (2) Interest charges under the condition that interest in any time period is only charged on the principal. Frequently, interest is charged on the original principal amount disregarding the fact that the principal still owing may be declining through time. (See Interest Rate, Nominal.)

Sinking Fund: (1) A fund accumulated by periodic deposits and reserved exclusively for a specific purpose, such as retirement of a debt or replacement of a property. (2) A fund created by making periodic deposits (usually equal) at compound interest in order to accumulate a given sum at a given future time usually for some specific purpose.

Sinking Fund Deposit Factor: (See Sinking Fund Factor.)

Sinking Fund Factor: The function of interest rate and time that determines the periodic deposit required to accumulate a specified future amount.

Study Period: The length of time that is presumed to be covered in the schedule of events and appraisal of results. Often, it is the anticipated life of the project under consideration, but may be either longer or (more likely) shorter. (See Life and Planning Horizon.)

Sunk Cost: A cost that, because it occurred in the past, has no relevance with respect to estimates of future receipts or disbursements. This concept implies that, because a past outlay is the same regardless of the alternative selected, it should not influence a new choice among alternatives.

Time Value of Money: (1) The cumulative effect of elapsed time and the money value of an event, based on the earning power of equivalent invested funds and on changes in purchasing power. (2) The expected compound interest rate that capital should or will earn. (See Interest.)

Traceable Costs: Cost elements that can be identified with a given product, operation, or service. (See Direct Cost and Marginal Cost.)

Uncertainty: (1) That which is indeterminate, indefinite, or problematical. (2) An attribute of the precision of an individual's or group's precision of knowledge that is about some fact, event, consequence, or measurement.

Uniform Gradient Series: A uniform or arithmetic pattern of receipts or disbursements that is increasing or decreasing by a constant amount in each time period. (See Gradient Factors.)

Utility: (1) In economics, a process of evaluating factor inputs and outputs in quantitative units (i.e., utiles) in order to arrive at a single measure of performance to assist in decision making. (2) In economic analysis, a measured preference among various choices available in risk situations based on the decision-making environment, the alternatives being considered, and the decision maker's personal attitudes.

Utility Function: A mathematically derived relationship between utility, measured in utiles, and quantities of money and/or commodities or attributes based on a decision maker's attitudes and preferences.

Valuation or Appraisal: The art and science of estimating the fair-exchange monetary value of specific properties.

Variable Cost: A cost that tends to fluctuate according to changes in the number of units produced. (Also see Marginal Cost.)

Working Capital: (1) The portion of investment that is represented by current assets (assets that are not capitalized) less the current liabilities. The capital that is necessary to sustain operations as opposed to that invested in fixed assets. (2) Those funds, other than investments in fixed assets, required to make the enterprise or project a going concern.

Yield: In evaluating investments, especially those offered by lending institutions, the true annual rate of return to the investor. In bond valuation, the annual dividend of a bond divided by the current market price and usually expressed as a percent. (See Expected Yield.)

appendix B

Engineering conversion factors

Science constants

speed of light	$2.997,925 \times 10^{10}$ cm/sec
	983.6×10^6 ft/sec
	186,284 miles/sec
velocity of sound	340.3 meters/sec
	1116 ft/sec
gravity	9.80665 m/sec square
(acceleration)	32.174 ft/sec square
	386.089 inches/sec square

Numbers and prefixes

yotta (10^{24}):	1 000 000 000 000 000 000 000 000
zetta (10^{21}):	1 000 000 000 000 000 000 000
exa (10^{18}):	1 000 000 000 000 000 000
peta (10^{15}):	1 000 000 000 000 000
tera (10^{12}):	1 000 000 000 000
giga (10^{9}):	1 000 000 000
mega (10^{6}):	1 000 000
kilo (10^{3}):	1 000
hecto (10^{2}):	100
deca (10^{1}):	10
deci (10^{-1}):	0.1
centi (10^{-2}):	0.01
milli (10^{-3}):	0.001
micro (10^{-6}):	0.000 001
nano (10^{-9}):	0.000 000 001
pico (10^{-12}):	0.000 000 000 001
femto (10^{-15}):	0.000 000 000 000 001
atto (10^{-18}):	0.000 000 000 000 000 001
zepto (10^{-21}):	0.000 000 000 000 000 000 001
yacto (10^{-24}):	0.000 000 000 000 000 000 000 001
Stringo (10^{-35}):	0.000 000 000 000 000 000 000 000 000 000 000 01

Area conversion factors

Multiply	by	to obtain
acres	43,560	sq feet
	4,047	sq meters
	4,840	sq yards
	0.405	hectare
sq cm	0.155	sq inches
sq feet	144	sq inches
	0.09290	sq meters
	0.1111	sq yards
sq inches	645.16	sq millimeters
sq kilometers	0.3861	sq miles
sq meters	10.764	sq feet
	1.196	sq yards
sq miles	640	acres
	2.590	sq kilometers

Volume conversion factors

Multiply	by	to obtain
acre-foot	1233.5	cubic meters
cubic cm	0.06102	cubic inches
cubic feet	1728	cubic inches
	7.480	gallons (U.S.)
	0.02832	cubic meters
	0.03704	cubic yards
liter	1.057	liquid quarts
	0.908	dry quarts
	61.024	cubic inches
gallons (U.S.)	231	cubic inches
	3.7854	liters
	4	quarts
	0.833	British gallons
	128	U.S. fluid ounces
quarts (U.S.)	0.9463	liters

Energy conversion factors

Multiply	by	to obtain
Btu	1055.9	joules
	0.2520	kg-calories
watt-hour	3600	joules
	3.409	Btu
HP (electric)	746	watts
Btu/second	1055.9	watts
watt-second	1.00	joules

Mass conversion factors

Multiply	by	to obtain
carat	0.200	cubic grams
grams	0.03527	ounces
kilograms	2.2046	pounds
ounces	28.350	grams
pound	16	ounces
	453.6	grams
stone (U.K.)	6.35	kilograms
	14	pounds
ton (net)	907.2	kilograms
	2000	pounds
	0.893	gross ton
	0.907	metric ton
ton (gross)	2240	pounds
	1.12	net tons
	1.016	metric tons
tonne (metric)	2,204.623	pounds
	0.984	gross pound
	1000	kilograms

Temperature conversion factors

Conversion formulas

Celsius to Kelvin	$K = C + 273.15$
Celsius to Fahrenheit	$F = (9/5)C + 32$
Fahrenheit to Celsius	$C = (5/9)(F - 32)$
Fahrenheit to Kelvin	$K = (5/9)(F + 459.67)$
Fahrenheit to Rankin	$R = F + 459.67$
Rankin to Kelvin	$K = (5/9)R$

Velocity conversion factors

Multiply	by	to obtain
feet/minute	5.080	mm/second
feet/second	0.3048	meters/second
inches/second	0.0254	meters/second
km/hour	0.6214	miles/hour
meters/second	3.2808	feet/second
	2.237	miles/hour
miles/hour	88.0	feet/minute
	0.44704	meters/second
	1.6093	km/hour
	0.8684	knots
knot	1.151	miles/hour

Pressure conversion factors

Multiply	by	to obtain
atmospheres	1.01325	bars
	33.90	feet of water
	29.92	inches of mercury
	760.0	mm of mercury
bar	75.01	cm of mercury
	14.50	pounds/sq inch
dyne/sq cm	0.1	N/sq meter
newtons/sq cm	1.450	pounds/sq inch
pounds/sq inch	0.06805	atmospheres
	2.036	inches of mercury
	27.708	inches of water
	68.948	millibars
	51.72	mm of mercury

Distance conversion factors

Multiply	by	to obtain
angstrom	10^{-10}	meters
feet	0.30480	meters
	12	inches
inches	25.40	millimeters
	0.02540	meters
	0.08333	feet
kilometers	3280.8	feet
	0.6214	miles
	1094	yards
meters	39.370	inches
	3.2808	feet
	1.094	yards
miles	5280	feet
	1.6093	kilometers
	0.8694	nautical miles
millimeters	0.03937	inches
nautical miles	6076	feet
	1.852	kilometers
yards	0.9144	meters
	3	feet
	36	inches

Physical science equations

$D = \dfrac{m}{V}$	D density m mass V volume	$\left(\dfrac{g}{cm^3} = \dfrac{kg}{m^3}\right)$	$P = \dfrac{W}{t}$	P power W(=watts) W work J t time s	
$d = v \cdot t$	d distance m v velocity m/s t time s		$K.E. = \dfrac{1}{2} \cdot m \cdot v^2$	K.E. kinetic energy m mass kg v velocity m/s	
$a = \dfrac{vf - vi}{t}$	a acceleration m/s² vf final velocity m/s vi initial velocity m/s t time s		$Fe = \dfrac{k \cdot Q_1 \cdot Q_2}{d^2}$	Fe electrical force N k Coulomb's constant $\left(k = 9 \times 10^9 \dfrac{N \cdot m^2}{c^2}\right)$	
$d = vi \cdot t + \dfrac{1}{2} \cdot a \cdot t^2$	d distance m vi initial velocity m/s t time s a acceleration m/s²			$Q_1 \cdot Q_2$ are electrical charges C d separation distance m	

$F = m \cdot a$	F net force N(=newtons)
	m mass kg
	a acceleration m/s²
			$V = \dfrac{W}{Q}$	V electrical potential
				   difference V(=volts)
				W work done J
				Q electric charge
				   moving C	

$Fg = \dfrac{G \cdot m_1 \cdot m_2}{d^2}$	Fg force of gravity N
	G universal
	   gravitational constant
	
	$\left(G = 6.67 \times 10^{-11} \dfrac{N - m^2}{kg^2}\right)$
			$I = \dfrac{Q}{t}$	I electric current amperes
				Q electric charge
				   flowing C
				t time s	

	m_1, m_2 masses of the two objects kg d separation distance m		$W = V \cdot I \cdot t$	W electrical energy J V voltage V I current A t time s	
$p = m \cdot v$	p momentum kg·m/s m mass v velocity		$P = V \cdot I$	P power W V voltage V I current A	
$W = F \cdot d$	W work J(=joules) F force N d distance m		$H = c \cdot m \cdot \Delta T$	H heat energy J m mass kg T change in temperature °C c specific heat J/Kg·°C	

Units of measurement

English system		Metric system		
1 foot (ft)	= 12 inches (in) 1'=12"	mm	millimeter	.001 m
1 yard (yd)	= 3 feet	cm	centimeter	.01 m
1 mile (mi)	= 1760 yards	dm	decimeter	.1 m
1 sq. foot	= 144 sq. inches	m	meter	1 m
1 sq. yard	= 9 sq. feet	dam	dekameter	10 m
1 acre	= 4840 sq. yards = 43,560 ft²	hm	hectometer	100 m
1 sq. mile	= 640 acres	km	kilometer	1000 m

Note: Prefixes also apply to l (liter) and g (gram).

Common Notations

Measurement	Notation	Description
meter	m	length
hectare	ha	area
tonne	t	mass
kilogram	kg	mass
nautical mile	M	distance (navigation)
knot	kn	speed (navigation)
liter	L	volume or capacity
second	s	time
hertz	Hz	frequency
candela	cd	luminous intensity
degree Celsius	°C	temperature
kelvin	K	thermodynamic temp.
pascal	Pa	pressure, stress
joule	J	energy, work
newton	N	force
watt	W	power, radiant flux
ampere	A	electric current
volt	V	electric potential
ohm	Ω	electric resistance
coulomb	C	electric charge

Household Measurements

A pinch	1/8 tsp. or less
3 tsp.	1 tbsp.
2 tbsp.	1/8 c.
4 tbsp.	1/4 c.
16 tbsp.	1 c.
5 tbsp. + 1 tsp.	1/3 c.
4 oz.	1/2 c.
8 oz.	1 c.
16 oz.	1 lb.
1 oz.	2 tbsp. fat or liquid
1 c. of liquid	1/2 pt.
2 c.	1 pt.
2 pt.	1 qt.
4 c. of liquid	1 qt.
4 qts.	1 gallon
8 qts.	1 peck (such as apples, pears, etc.)
1 jigger	1 ∫ fl.oz.
1 jigger	3 tbsp.

Computational and mathematical formulae

$$\sum_{n=0}^{\infty} \frac{x^n}{n!} = e^x$$

$$\sum_{n=0}^{\infty} \frac{x^n}{n} = \ln\left(\frac{1}{1-x}\right)$$

$$\sum_{n=0}^{k} x^n = \frac{x^{k+1}-1}{x-1}, \qquad x \neq 1$$

$$\sum_{n=1}^{k} x^n = \frac{x-x^{k+1}}{1-x}, \qquad x \neq 1$$

$$\sum_{n=2}^{k} x^n = \frac{x^2-x^{k+1}}{1-x}, \qquad x \neq 1$$

$$\sum_{n=0}^{\infty} p^n = \frac{1}{1-p}, \qquad \text{if } |p|<1$$

$$\sum_{n=0}^{\infty} nx^n = \frac{x}{(1-x)^2}, \qquad x \neq 1$$

$$\sum_{n=0}^{\infty} n^2 x^n = \frac{2x^2}{(1-x)^3} + \frac{x}{(1-x)^2}, \quad x \neq 1$$

$$\sum_{n=0}^{\infty} n^3 x^n = \frac{6x^3}{(1-x)^4} + \frac{6x^2}{(1-x)^3} + \frac{x}{(1-x)^2}, \quad x \neq 1$$

$$\sum_{n=0}^{M} n x^n = \frac{x\left[1 - (M+1)x^M + Mx^{M+1}\right]}{(1-x)^2}, \quad x \neq 1$$

$$\sum_{x=0}^{\infty} \binom{r+x-1}{x} u^x = (1-u)^{-r}, \quad \text{if } |u| < 1$$

$$\sum_{k=1}^{\infty} (-1)^{k+1} \frac{1}{k} = 1 - \frac{1}{2} + \frac{1}{3} - \frac{1}{4} + \frac{1}{5} - \frac{1}{6} + \dots = \ln 2$$

$$\sum_{k=1}^{\infty} (-1)^{k+1} \frac{1}{(2k-1)} = 1 - \frac{1}{3} + \frac{1}{5} - \frac{1}{7} + \frac{1}{9} - \dots = \frac{\pi}{4}$$

$$\sum_{k=0}^{\infty} (-1)^k x^k = \frac{1}{1+x}, \quad -1 < x < 1$$

$$\sum_{k=1}^{n} (-1)^k \binom{n}{k} = 1, \quad \text{for } n \geq 2$$

$$\sum_{k=0}^{n} \binom{n}{k}^2 = \binom{2n}{n}$$

$$\sum_{k=1}^{n} k = 1 + 2 + 3 + \dots + n = \frac{n(n+1)}{2}$$

$$\sum_{k=1}^{n} (2k) = 2 + 4 + 6 + \dots + 2n = n(n-1)$$

$$\sum_{k=1}^{n}\left(2k-1\right)=1+3+5+...+\left(2n-1\right)=n^2$$

$$\sum_{k=0}^{\infty}\left(a+kd\right)r^k=a+\left(a+d\right)r+\left(a+2d\right)r^2+...+=\frac{a}{1-r}+\frac{rd}{\left(1-r\right)^2}$$

$$\sum_{k=1}^{n}k^2=1+4+9+...+n^2=\frac{n\left(n+1\right)\left(2n+1\right)}{6}$$

$$\sum_{k=1}^{n}k^3=1+8+27+...+n^3=\frac{n^2\left(n+1\right)^2}{4}=\left[\frac{n\left(n+1\right)}{2}\right]^2=\left[\sum_{k=1}^{n}k\right]^2$$

$$\sum_{x=1}^{\infty}\frac{1}{x}=1+\frac{1}{2}+\frac{1}{3}+...\text{(does not converge)}$$

$$\sum_{m=0}^{k}ma^m=\frac{a}{\left(1-a\right)^2}\left[1-\left(k+1\right)a^k+ka^{k+1}\right]=\sum_{m=1}^{k}ma^m$$

$$\sum_{k=0}^{n}\left(1\right)=n$$

$$\sum_{k=0}^{n}\binom{n}{k}=2^n$$

$$\left(a+b\right)^n=\sum_{k=0}^{n}\binom{n}{k}a^kb^{n-k}$$

$$\prod_{n=1}^{\infty}a_n=e^{\left(\sum_{n=1}^{\infty}\ln a_n\right)}$$

$$\ln\left(\prod_{n=1}^{\infty}a_n\right)=\sum_{n=1}^{\infty}\ln a_n$$

$$\ln(x) = \sum_{k=1}^{\infty} \frac{1}{k}\left(\frac{x-1}{x}\right)^k, \quad x \geq \frac{1}{2}$$

$$\lim_{h \to \infty}(1+h)^{1/h} = e$$

$$\lim_{n \to \infty}\left(1+\frac{x}{n}\right)^n = e^{-x}$$

$$\lim_{n \to \infty}\sum_{k=0}^{n}\frac{e^{-n}n^r}{K!} = \frac{1}{2}$$

$$\lim_{k \to \infty}\left(\frac{x^k}{k!}\right) = 0$$

$$|x+y| \leq |x| + |y|$$

$$|x-y| \geq |x| - |y|$$

$$\ln(1+x) = \sum_{k=1}^{\infty}(-1)^{k+1}\left(\frac{x^k}{k}\right), \quad \text{if } -1 < x \leq 1$$

$$\Gamma\left(\frac{1}{2}\right) = \sqrt{\pi}$$

$$\Gamma(\alpha + 1) = \alpha\Gamma(\alpha)$$

$$\Gamma\left(\frac{n}{2}\right) = \frac{\sqrt{\pi}(n-1)!}{2^{n-1}\left(\frac{n-1}{2}\right)!}, \quad n \text{ odd}$$

$$\int_0^{\infty} e^{-x}x^{n-1}dx = \Gamma(n)$$

$$\binom{n}{2} = \frac{1}{2}\left(n^2 - n\right) = \sum_{k=1}^{n-1} K$$

$$\binom{n+1}{2} = \binom{n}{2} + n$$

$$2.4.6.8...2n = \prod_{k=1}^{n} 2k = 2^n n!$$

$$1.3.5.7...\left(2n - 1\right) = \frac{\left(2n - 1\right)!}{2^{2n-2}\left(2n-2\right)!} = \frac{2n - 1}{2^{2n-2}}$$

Derivation of closed form expression for $\sum_{k=1}^{n} kx^k$:

$$\sum_{k=1}^{n} kx^k = x \sum_{k=1}^{n} kx^{k-1}$$

$$= x \sum_{k=1}^{n} \frac{d}{dx}\left[x^k\right]$$

$$= x \frac{d}{dx}\left[\sum_{k=1}^{n} x^k\right]$$

$$= x \frac{d}{dx}\left[\frac{x\left(1-x^n\right)}{1-x}\right]$$

$$= x \left[\frac{\left(1-\left(n+1\right)x^n\right)\left(1-x\right) - x\left(1-x^n\right)\left(-1\right)}{\left(1-x\right)^2}\right]$$

$$= \frac{x\left[1 - \left(n + 1\right)x^n + nx^{n+1}\right]}{\left(1 - x\right)^2}, \quad x \neq 1$$

appendix D

Units of measure

Acre: An area of 43,560 square feet.

Agate: 1/14 inch (used in printing for measuring column length).

Ampere: Unit of electric current.

Astronomical (A.U.): 93,000,000 miles; the average distance of the Earth from the sun (used in astronomy).

Bale: A large bundle of goods. In the United States, the approximate weight of a bale of cotton is 500 pounds. Weight of a bale may vary from country to country.

Board Foot: 144 cubic inches (12 by 12 by 1 used for lumber).

Bolt: 40 yards (used for measuring cloth).

Btu: British thermal unit; amount of heat needed to increase the temperature of one pound of water by one degree Fahrenheit (252 calories).

Carat: 200 milligrams or 3,086 troy; used for weighing precious stones (originally the weight of a seed of the carob tree in the Mediterranean region). *See also Karat.*

Chain: 66 feet; used in surveying (one mile = 80 chains).

Cubit: 18 inches (derived from the distance between the elbow and the tip of the middle finger).

Decibel: Unit of relative loudness.

Freight Ton: 40 cubic feet of merchandise (used for cargo freight).

Gross: 12 dozen (144).

Hertz: Unit of measurement of electromagnetic wave frequencies (measures cycles per second).

Hogshead: 2 liquid barrels or 14,653 cubic inches.

Horsepower: The power needed to lift 33,000 pounds a distance of one foot in one minute (about 1½ times the power that an average horse can exert); used for measuring the power of mechanical engines.

Karat: A measure of the purity of gold. It indicates how many parts out of 24 are pure. 18 karat gold is Ω pure gold.

Knot: Rate of speed of 1 nautical mile per hour; used for measuring speed of ships (not distance).

League: Approximately 3 miles.

Light year: 5,880,000,000,000 miles; distance traveled by light in one year at the rate of 186,281.7 miles per second; used for measurement of interstellar space.

Magnum: Two-quart bottle; used for measuring wine.

Ohm: Unit of electrical resistance.

Parsec: Approximately 3.26 light years of 19.2 trillion miles; used for measuring interstellar distances.

Pi (π): 3.14159265+; the ratio of the circumference of a circle to its diameter.

Pica: 1/6 inch or 12 points; used in printing for measuring column width.

Pipe: 2 hogsheads; used for measuring wine and other liquids.

Point: 0.013837 (approximately 1/72 inch or 1/12 pica); used in printing for measuring type size.

Quintal: 100,000 grams or 220.46 pounds avoirdupois.

Quire: 24 or 25 sheets; used for measuring paper (20 quires is one ream).

Ream: 480 or 500 sheets; used for measuring paper.

Roentgen: Dosage unit of radiation exposure produced by x-rays.

Score: 20 units.

Span: 9 inches or 22.86 cm; derived from the distance between the end of the thumb and the end of the little finger when both are outstretched.

Square: 100 square feet; used in building.

Stone: 14 pounds avoirdupois in Great Britain.

Therm: 100,000 Btus.

Township: U.S. land measurement of almost 36 square miles; used in surveying.

Tun: 252 gallons (sometimes larger); used for measuring wine and other liquids.

Watt: Unit of power.

appendix E

Interest factors and tables

Formulas for interest factor

Name of Factor	Formula	Table Notation
Compound Amount (single payment)	$(1+i)^N$	(F/P, i, N)
Present Worth (single payment)	$(1+i)^{-N}$	(P/F, i, N)
Sinking Fund	$\dfrac{i}{(1+i)^N - 1}$	(A/F, i, N)
Capital Recovery	$\dfrac{i(1+i)^N}{(1+i)^N - 1}$	(A/P, i, N)
Compound Amount (uniform series)	$\dfrac{(1+i)^N - 1}{i}$	(F/A, i, N)
Present Worth (uniform series)	$\dfrac{(1+i)^N - 1}{i(1+i)^N}$	(P/A, i, N)
Arithmetic Gradient to Uniform Series	$\dfrac{(1+i)^N - iN - 1}{i(1+i)^N - i}$	(A/G, i, N)
Arithmetic Gradient to Present Worth	$\dfrac{(1+i)^N - iN - 1}{i^2(1+i)^N}$	(P/G, i, N)
Geometric Gradient to Present Worth (for ig)	$\dfrac{1-(1+g)^N(1+i)^{-N}}{i-g}$	(P/A, g, i, N)

Continuous Compounding Compound Amount (single payment)	e^{rN}	(F/P, r, N)
Continuous Compounding Present Worth (single payment)	e^{-rN}	(P/F, r, N)
Continuous Compounding Present Worth (uniform series)	$\dfrac{e^{rN}-1}{e^{rN}(e^{r}-1)}$	(P/A, r, N)
Continuous Compounding Sinking Fund	$\dfrac{e^{r}-1}{e^{rN}-1}$	(A/F, r, N)
Continuous Compounding Capital Recovery	$\dfrac{e^{rN}(e^{r}-1)}{e^{rN}-1}$	(A/P, r, N)
Continuous Compounding Compound Amount (uniform series)	$\dfrac{e^{rN}-1}{e^{r}-1}$	(F/A, r, N)
Continuous Compounding Present Worth (single, continuous payment)	$\dfrac{i(1+i)^{-N}}{\ln(1+i)}$	(P/$\overline{\text{F}}$, i, N)
Continuous Compounding Compound Amount (single, continuous payment)	$\dfrac{i(1+i)^{N-1}}{\ln(1+i)}$	(F/$\overline{\text{P}}$, i, N)
Continuous Compounding Sinking Fund (continuous, uniform payments)	$\dfrac{\ln(1+i)}{(1+i)^{N}-1}$	($\overline{\text{A}}$/F, i, N)
Continuous Compounding Capital Recovery (continuous, uniform payments)	$\dfrac{(1+i)^{N}\ln(1+i)}{(1+i)^{N}-1}$	($\overline{\text{A}}$/P, i, N)
Continuous Compounding Compound Amount (continuous, uniform payments)	$\dfrac{(1+i)^{N}-1}{\ln(1+i)}$	(F/$\overline{\text{A}}$, i, N)
Continuous Compounding Present Worth (continuous, uniform payments)	$\dfrac{(1+i)^{N}-1}{(1+i)^{N}\ln(1+i)}$	(P/$\overline{\text{A}}$, i, N)

Summation formulas for closed-form expressions

$$\sum_{t=0}^{n} x^t = \frac{1 - x^{n+1}}{1 - x}$$

$$\sum_{t=0}^{n-1} x^t = \frac{1 - x^{n}}{1 - x}$$

$$\sum_{t=1}^{n} x^t = \frac{x - x^{n+1}}{1 - x}$$

$$\sum_{t=1}^{n-1} x^t = \frac{x - x^{n}}{1 - x}$$

$$\sum_{t=2}^{n} x^t = \frac{x^2 - x^{n+1}}{1 - x}$$

$$\sum_{t=0}^{n} x^t = \frac{x\left[1 - (n+1)x^{n} + nx^{n+1}\right]}{(1-x)^2} = \frac{x - (1-x)(n+1)x^{n+1} - x^{n+2}}{(1-x)^2}$$

$$\sum_{t=1}^{n} tx^t = \sum_{t=0}^{n} tx^t - 0x^0 = \sum_{t=0}^{n} tx^t - 0 = \sum_{t=0}^{n} tx^t$$

$$= \frac{x\left[1 - (n+1)x^{n} + nx^{n+1}\right]}{(1-x)^2} = \frac{x - (1-x)(n+1)x^{n+1} - x^{n+2}}{(1-x)^2}$$

Interest tables

0.25%				Compound Interest Factors					0.25%
Period	Single Payment		Uniform Payment Series				Arithmetic Gradient		Period
	Compound Amount Factor	Present Value Factor	Sinking Fund Factor	Capital Recovery Factor	Compound Amount Factor	Present Value Factor	Gradient Uniform Series	Gradient Present Value	
	Find F Given P	Find P Given F	Find A Given F	Find A Given P	Find F Given A	Find P Given A	Find A Given G	Find P Given G	
n	F/P	P/F	A/F	A/P	F/A	P/A	A/G	P/G	n
1	1.003	0.9975	1.0000	1.0025	1.000	0.998	0.000	0.000	1
2	1.005	0.9950	0.4994	0.5019	2.002	1.993	0.499	0.995	2
3	1.008	0.9925	0.3325	0.3350	3.008	2.985	0.998	2.980	3
4	1.010	0.9901	0.2491	0.2516	4.015	3.975	1.497	5.950	4
5	1.013	0.9876	0.1990	0.2015	5.025	4.963	1.995	9.901	5
6	1.015	0.9851	0.1656	0.1681	6.038	5.948	2.493	14.826	6
7	1.018	0.9827	0.1418	0.1443	7.053	6.931	2.990	20.722	7
8	1.020	0.9802	0.1239	0.1264	8.070	7.911	3.487	27.584	8
9	1.023	0.9778	0.1100	0.1125	9.091	8.889	3.983	35.406	9
10	1.025	0.9753	0.0989	0.1014	10.113	9.864	4.479	44.184	10
11	1.028	0.9729	0.0898	0.0923	11.139	10.837	4.975	53.913	11
12	1.030	0.9705	0.0822	0.0847	12.166	11.807	5.470	64.589	12
13	1.033	0.9681	0.0758	0.0783	13.197	12.775	5.965	76.205	13
14	1.036	0.9656	0.0703	0.0728	14.230	13.741	6.459	88.759	14
15	1.038	0.9632	0.0655	0.0680	15.265	14.704	6.953	102.244	15
16	1.041	0.9608	0.0613	0.0638	16.304	15.665	7.447	116.657	16
17	1.043	0.9584	0.0577	0.0602	17.344	16.623	7.940	131.992	17
18	1.046	0.9561	0.0544	0.0569	18.388	17.580	8.433	148.245	18
19	1.049	0.9537	0.0515	0.0540	19.434	18.533	8.925	165.411	19
20	1.051	0.9513	0.0488	0.0513	20.482	19.484	9.417	183.485	20
21	1.054	0.9489	0.0464	0.0489	21.533	20.433	9.908	202.463	21
22	1.056	0.9466	0.0443	0.0468	22.587	21.380	10.400	222.341	22
23	1.059	0.9442	0.0423	0.0448	23.644	22.324	10.890	243.113	23
24	1.062	0.9418	0.0405	0.0430	24.703	23.266	11.380	264.775	24
25	1.064	0.9395	0.0388	0.0413	25.765	24.205	11.870	287.323	25
26	1.067	0.9371	0.0373	0.0398	26.829	25.143	12.360	310.752	26
27	1.070	0.9348	0.0358	0.0383	27.896	26.077	12.849	335.057	27
28	1.072	0.9325	0.0345	0.0370	28.966	27.010	13.337	360.233	28
29	1.075	0.9301	0.0333	0.0358	30.038	27.940	13.825	386.278	29
30	1.078	0.9278	0.0321	0.0346	31.113	28.868	14.313	413.185	30
31	1.080	0.9255	0.0311	0.0336	32.191	29.793	14.800	440.950	31
32	1.083	0.9232	0.0301	0.0326	33.272	30.717	15.287	469.570	32
33	1.086	0.9209	0.0291	0.0316	34.355	31.638	15.774	499.039	33
34	1.089	0.9186	0.0282	0.0307	35.441	32.556	16.260	529.353	34
35	1.091	0.9163	0.0274	0.0299	36.529	33.472	16.745	560.508	35
36	1.094	0.9140	0.0266	0.0291	37.621	34.386	17.231	592.499	36
37	1.097	0.9118	0.0258	0.0283	38.715	35.298	17.715	625.322	37
38	1.100	0.9095	0.0251	0.0276	39.811	36.208	18.200	658.973	38
39	1.102	0.9072	0.0244	0.0269	40.911	37.115	18.684	693.447	39
40	1.105	0.9050	0.0238	0.0263	42.013	38.020	19.167	728.740	40
41	1.108	0.9027	0.0232	0.0257	43.118	38.923	19.650	764.848	41
42	1.111	0.9004	0.0226	0.0251	44.226	39.823	20.133	801.766	42
43	1.113	0.8982	0.0221	0.0246	45.337	40.721	20.616	839.490	43
44	1.116	0.8960	0.0215	0.0240	46.450	41.617	21.097	878.016	44
45	1.119	0.8937	0.0210	0.0235	47.566	42.511	21.579	917.340	45
46	1.122	0.8915	0.0205	0.0230	48.685	43.402	22.060	957.457	46
47	1.125	0.8893	0.0201	0.0226	49.807	44.292	22.541	998.364	47
48	1.127	0.8871	0.0196	0.0221	50.931	45.179	23.021	1040.055	48
49	1.130	0.8848	0.0192	0.0217	52.059	46.064	23.501	1082.528	49
50	1.133	0.8826	0.0188	0.0213	53.189	46.946	23.980	1125.777	50

0.5%				Compound Interest Factors					0.5%
Period	Single Payment		Uniform Payment Series				Arithmetic Gradient		Period
	Compound Amount Factor	Present Value Factor	Sinking Fund Factor	Capital Recovery Factor	Compound Amount Factor	Present Value Factor	Gradient Uniform Series	Gradient Present Value	
	Find F Given P	Find P Given F	Find A Given F	Find A Given P	Find F Given A	Find P Given A	Find A Given G	Find P Given G	
n	F/P	P/F	A/F	A/P	F/A	P/A	A/G	P/G	n
1	1.005	0.9950	1.0000	1.0050	1.000	0.995	0.000	0.000	1
2	1.010	0.9901	0.4988	0.5038	2.005	1.985	0.499	0.990	2
3	1.015	0.9851	0.3317	0.3367	3.015	2.970	0.997	2.960	3
4	1.020	0.9802	0.2481	0.2531	4.030	3.950	1.494	5.901	4
5	1.025	0.9754	0.1980	0.2030	5.050	4.926	1.990	9.803	5
6	1.030	0.9705	0.1646	0.1696	6.076	5.896	2.485	14.655	6
7	1.036	0.9657	0.1407	0.1457	7.106	6.862	2.980	20.449	7
8	1.041	0.9609	0.1228	0.1278	8.141	7.823	3.474	27.176	8
9	1.046	0.9561	0.1089	0.1139	9.182	8.779	3.967	34.824	9
10	1.051	0.9513	0.0978	0.1028	10.228	9.730	4.459	43.386	10
11	1.056	0.9466	0.0887	0.0937	11.279	10.677	4.950	52.853	11
12	1.062	0.9419	0.0811	0.0861	12.336	11.619	5.441	63.214	12
13	1.067	0.9372	0.0746	0.0796	13.397	12.556	5.930	74.460	13
14	1.072	0.9326	0.0691	0.0741	14.464	13.489	6.419	86.583	14
15	1.078	0.9279	0.0644	0.0694	15.537	14.417	6.907	99.574	15
16	1.083	0.9233	0.0602	0.0652	16.614	15.340	7.394	113.424	16
17	1.088	0.9187	0.0565	0.0615	17.697	16.259	7.880	128.123	17
18	1.094	0.9141	0.0532	0.0582	18.786	17.173	8.366	143.663	18
19	1.099	0.9096	0.0503	0.0553	19.880	18.082	8.850	160.036	19
20	1.105	0.9051	0.0477	0.0527	20.979	18.987	9.334	177.232	20
21	1.110	0.9006	0.0453	0.0503	22.084	19.888	9.817	195.243	21
22	1.116	0.8961	0.0431	0.0481	23.194	20.784	10.299	214.061	22
23	1.122	0.8916	0.0411	0.0461	24.310	21.676	10.781	233.677	23
24	1.127	0.8872	0.0393	0.0443	25.432	22.563	11.261	254.082	24
25	1.133	0.8828	0.0377	0.0427	26.559	23.446	11.741	275.269	25
26	1.138	0.8784	0.0361	0.0411	27.692	24.324	12.220	297.228	26
27	1.144	0.8740	0.0347	0.0397	28.830	25.198	12.698	319.952	27
28	1.150	0.8697	0.0334	0.0384	29.975	26.068	13.175	343.433	28
29	1.156	0.8653	0.0321	0.0371	31.124	26.933	13.651	367.663	29
30	1.161	0.8610	0.0310	0.0360	32.280	27.794	14.126	392.632	30
31	1.167	0.8567	0.0299	0.0349	33.441	28.651	14.601	418.335	31
32	1.173	0.8525	0.0289	0.0339	34.609	29.503	15.075	444.762	32
33	1.179	0.8482	0.0279	0.0329	35.782	30.352	15.548	471.906	33
34	1.185	0.8440	0.0271	0.0321	36.961	31.196	16.020	499.758	34
35	1.191	0.8398	0.0262	0.0312	38.145	32.035	16.492	528.312	35
36	1.197	0.8356	0.0254	0.0304	39.336	32.871	16.962	557.560	36
37	1.203	0.8315	0.0247	0.0297	40.533	33.703	17.432	587.493	37
38	1.209	0.8274	0.0240	0.0290	41.735	34.530	17.901	618.105	38
39	1.215	0.8232	0.0233	0.0283	42.944	35.353	18.369	649.388	39
40	1.221	0.8191	0.0226	0.0276	44.159	36.172	18.836	681.335	40
41	1.227	0.8151	0.0220	0.0270	45.380	36.987	19.302	713.937	41
42	1.233	0.8110	0.0215	0.0265	46.607	37.798	19.768	747.189	42
43	1.239	0.8070	0.0209	0.0259	47.840	38.605	20.233	781.081	43
44	1.245	0.8030	0.0204	0.0254	49.079	39.408	20.696	815.609	44
45	1.252	0.7990	0.0199	0.0249	50.324	40.207	21.159	850.763	45
46	1.258	0.7950	0.0194	0.0244	51.576	41.002	21.622	886.538	46
47	1.264	0.7910	0.0189	0.0239	52.834	41.793	22.083	922.925	47
48	1.270	0.7871	0.0185	0.0235	54.098	42.580	22.544	959.919	48
49	1.277	0.7832	0.0181	0.0231	55.368	43.364	23.003	997.512	49
50	1.283	0.7793	0.0177	0.0227	56.645	44.143	23.462	1035.697	50

0.75%				Compound Interest Factors					0.75%
Period	Single Payment		Uniform Payment Series				Arithmetic Gradient		Period
	Compound Amount Factor	Present Value Factor	Sinking Fund Factor	Capital Recovery Factor	Compound Amount Factor	Present Value Factor	Gradient Uniform Series	Gradient Present Value	
	Find F Given P	Find P Given F	Find A Given F	Find A Given P	Find F Given A	Find P Given A	Find A Given G	Find P Given G	
n	F/P	P/F	A/F	A/P	F/A	P/A	A/G	P/G	n
1	1.008	0.9926	1.0000	1.0075	1.000	0.993	0.000	0.000	1
2	1.015	0.9852	0.4981	0.5056	2.008	1.978	0.498	0.985	2
3	1.023	0.9778	0.3308	0.3383	3.023	2.956	0.995	2.941	3
4	1.030	0.9706	0.2472	0.2547	4.045	3.926	1.491	5.852	4
5	1.038	0.9633	0.1970	0.2045	5.076	4.889	1.985	9.706	5
6	1.046	0.9562	0.1636	0.1711	6.114	5.846	2.478	14.487	6
7	1.054	0.9490	0.1397	0.1472	7.159	6.795	2.970	20.181	7
8	1.062	0.9420	0.1218	0.1293	8.213	7.737	3.461	26.775	8
9	1.070	0.9350	0.1078	0.1153	9.275	8.672	3.950	34.254	9
10	1.078	0.9280	0.0967	0.1042	10.344	9.600	4.438	42.606	10
11	1.086	0.9211	0.0876	0.0951	11.422	10.521	4.925	51.817	11
12	1.094	0.9142	0.0800	0.0875	12.508	11.435	5.411	61.874	12
13	1.102	0.9074	0.0735	0.0810	13.601	12.342	5.895	72.763	13
14	1.110	0.9007	0.0680	0.0755	14.703	13.243	6.379	84.472	14
15	1.119	0.8940	0.0632	0.0707	15.814	14.137	6.861	96.988	15
16	1.127	0.8873	0.0591	0.0666	16.932	15.024	7.341	110.297	16
17	1.135	0.8807	0.0554	0.0629	18.059	15.905	7.821	124.389	17
18	1.144	0.8742	0.0521	0.0596	19.195	16.779	8.299	139.249	18
19	1.153	0.8676	0.0492	0.0567	20.339	17.647	8.776	154.867	19
20	1.161	0.8612	0.0465	0.0540	21.491	18.508	9.252	171.230	20
21	1.170	0.8548	0.0441	0.0516	22.652	19.363	9.726	188.325	21
22	1.179	0.8484	0.0420	0.0495	23.822	20.211	10.199	206.142	22
23	1.188	0.8421	0.0400	0.0475	25.001	21.053	10.671	224.668	23
24	1.196	0.8358	0.0382	0.0457	26.188	21.889	11.142	243.892	24
25	1.205	0.8296	0.0365	0.0440	27.385	22.719	11.612	263.803	25
26	1.214	0.8234	0.0350	0.0425	28.590	23.542	12.080	284.389	26
27	1.224	0.8173	0.0336	0.0411	29.805	24.359	12.547	305.639	27
28	1.233	0.8112	0.0322	0.0397	31.028	25.171	13.013	327.542	28
29	1.242	0.8052	0.0310	0.0385	32.261	25.976	13.477	350.087	29
30	1.251	0.7992	0.0298	0.0373	33.503	26.775	13.941	373.263	30
31	1.261	0.7932	0.0288	0.0363	34.754	27.568	14.403	397.060	31
32	1.270	0.7873	0.0278	0.0353	36.015	28.356	14.864	421.468	32
33	1.280	0.7815	0.0268	0.0343	37.285	29.137	15.323	446.475	33
34	1.289	0.7757	0.0259	0.0334	38.565	29.913	15.782	472.071	34
35	1.299	0.7699	0.0251	0.0326	39.854	30.683	16.239	498.247	35
36	1.309	0.7641	0.0243	0.0318	41.153	31.447	16.695	524.992	36
37	1.318	0.7585	0.0236	0.0311	42.461	32.205	17.149	552.297	37
38	1.328	0.7528	0.0228	0.0303	43.780	32.958	17.603	580.151	38
39	1.338	0.7472	0.0222	0.0297	45.108	33.705	18.055	608.545	39
40	1.348	0.7416	0.0215	0.0290	46.446	34.447	18.506	637.469	40
41	1.358	0.7361	0.0209	0.0284	47.795	35.183	18.956	666.914	41
42	1.369	0.7306	0.0203	0.0278	49.153	35.914	19.404	696.871	42
43	1.379	0.7252	0.0198	0.0273	50.522	36.639	19.851	727.330	43
44	1.389	0.7198	0.0193	0.0268	51.901	37.359	20.297	758.281	44
45	1.400	0.7145	0.0188	0.0263	53.290	38.073	20.742	789.717	45
46	1.410	0.7091	0.0183	0.0258	54.690	38.782	21.186	821.628	46
47	1.421	0.7039	0.0178	0.0253	56.100	39.486	21.628	854.006	47
48	1.431	0.6986	0.0174	0.0249	57.521	40.185	22.069	886.840	48
49	1.442	0.6934	0.0170	0.0245	58.952	40.878	22.509	920.124	49
50	1.453	0.6883	0.0166	0.0241	60.394	41.566	22.948	953.849	50

1%			Compound Interest Factors						1%
Period	Single Payment		Uniform Payment Series				Arithmetic Gradient		Period
	Compound Amount Factor	Present Value Factor	Sinking Fund Factor	Capital Recovery Factor	Compound Amount Factor	Present Value Factor	Gradient Uniform Series	Gradient Present Value	
	Find F Given P	Find P Given F	Find A Given F	Find A Given P	Find F Given A	Find P Given A	Find A Given G	Find P Given G	
n	F/P	P/F	A/F	A/P	F/A	P/A	A/G	P/G	n
1	1.010	0.9901	1.0000	1.0100	1.000	0.990	0.000	0.000	1
2	1.020	0.9803	0.4975	0.5075	2.010	1.970	0.498	0.980	2
3	1.030	0.9706	0.3300	0.3400	3.030	2.941	0.993	2.921	3
4	1.041	0.9610	0.2463	0.2563	4.060	3.902	1.488	5.804	4
5	1.051	0.9515	0.1960	0.2060	5.101	4.853	1.980	9.610	5
6	1.062	0.9420	0.1625	0.1725	6.152	5.795	2.471	14.321	6
7	1.072	0.9327	0.1386	0.1486	7.214	6.728	2.960	19.917	7
8	1.083	0.9235	0.1207	0.1307	8.286	7.652	3.448	26.381	8
9	1.094	0.9143	0.1067	0.1167	9.369	8.566	3.934	33.696	9
10	1.105	0.9053	0.0956	0.1056	10.462	9.471	4.418	41.843	10
11	1.116	0.8963	0.0865	0.0965	11.567	10.368	4.901	50.807	11
12	1.127	0.8874	0.0788	0.0888	12.683	11.255	5.381	60.569	12
13	1.138	0.8787	0.0724	0.0824	13.809	12.134	5.861	71.113	13
14	1.149	0.8700	0.0669	0.0769	14.947	13.004	6.338	82.422	14
15	1.161	0.8613	0.0621	0.0721	16.097	13.865	6.814	94.481	15
16	1.173	0.8528	0.0579	0.0679	17.258	14.718	7.289	107.273	16
17	1.184	0.8444	0.0543	0.0643	18.430	15.562	7.761	120.783	17
18	1.196	0.8360	0.0510	0.0610	19.615	16.398	8.232	134.996	18
19	1.208	0.8277	0.0481	0.0581	20.811	17.226	8.702	149.895	19
20	1.220	0.8195	0.0454	0.0554	22.019	18.046	9.169	165.466	20
21	1.232	0.8114	0.0430	0.0530	23.239	18.857	9.635	181.695	21
22	1.245	0.8034	0.0409	0.0509	24.472	19.660	10.100	198.566	22
23	1.257	0.7954	0.0389	0.0489	25.716	20.456	10.563	216.066	23
24	1.270	0.7876	0.0371	0.0471	26.973	21.243	11.024	234.180	24
25	1.282	0.7798	0.0354	0.0454	28.243	22.023	11.483	252.894	25
26	1.295	0.7720	0.0339	0.0439	29.526	22.795	11.941	272.196	26
27	1.308	0.7644	0.0324	0.0424	30.821	23.560	12.397	292.070	27
28	1.321	0.7568	0.0311	0.0411	32.129	24.316	12.852	312.505	28
29	1.335	0.7493	0.0299	0.0399	33.450	25.066	13.304	333.486	29
30	1.348	0.7419	0.0287	0.0387	34.785	25.808	13.756	355.002	30
31	1.361	0.7346	0.0277	0.0377	36.133	26.542	14.205	377.039	31
32	1.375	0.7273	0.0267	0.0367	37.494	27.270	14.653	399.586	32
33	1.389	0.7201	0.0257	0.0357	38.869	27.990	15.099	422.629	33
34	1.403	0.7130	0.0248	0.0348	40.258	28.703	15.544	446.157	34
35	1.417	0.7059	0.0240	0.0340	41.660	29.409	15.987	470.158	35
36	1.431	0.6989	0.0232	0.0332	43.077	30.108	16.428	494.621	36
37	1.445	0.6920	0.0225	0.0325	44.508	30.800	16.868	519.533	37
38	1.460	0.6852	0.0218	0.0318	45.953	31.485	17.306	544.884	38
39	1.474	0.6784	0.0211	0.0311	47.412	32.163	17.743	570.662	39
40	1.489	0.6717	0.0205	0.0305	48.886	32.835	18.178	596.856	40
41	1.504	0.6650	0.0199	0.0299	50.375	33.500	18.611	623.456	41
42	1.519	0.6584	0.0193	0.0293	51.879	34.158	19.042	650.451	42
43	1.534	0.6519	0.0187	0.0287	53.398	34.810	19.472	677.831	43
44	1.549	0.6454	0.0182	0.0282	54.932	35.455	19.901	705.585	44
45	1.565	0.6391	0.0177	0.0277	56.481	36.095	20.327	733.704	45
46	1.580	0.6327	0.0172	0.0272	58.046	36.727	20.752	762.176	46
47	1.596	0.6265	0.0168	0.0268	59.626	37.354	21.176	790.994	47
48	1.612	0.6203	0.0163	0.0263	61.223	37.974	21.598	820.146	48
49	1.628	0.6141	0.0159	0.0259	62.835	38.588	22.018	849.624	49
50	1.645	0.6080	0.0155	0.0255	64.463	39.196	22.436	879.418	50

1.25%			Compound Interest Factors						1.25%
Period	Single Payment		Uniform Payment Series				Arithmetic Gradient		Period
	Compound Amount Factor	Present Value Factor	Sinking Fund Factor	Capital Recovery Factor	Compound Amount Factor	Present Value Factor	Gradient Uniform Series	Gradient Present Value	
	Find F Given P	Find P Given F	Find A Given F	Find A Given P	Find F Given A	Find P Given A	Find A Given G	Find P Given G	
n	F/P	P/F	A/F	A/P	F/A	P/A	A/G	P/G	n
1	1.013	0.9877	1.0000	1.0125	1.000	0.988	0.000	0.000	1
2	1.025	0.9755	0.4969	0.5094	2.013	1.963	0.497	0.975	2
3	1.038	0.9634	0.3292	0.3417	3.038	2.927	0.992	2.902	3
4	1.051	0.9515	0.2454	0.2579	4.076	3.878	1.484	5.757	4
5	1.064	0.9398	0.1951	0.2076	5.127	4.818	1.975	9.516	5
6	1.077	0.9282	0.1615	0.1740	6.191	5.746	2.464	14.157	6
7	1.091	0.9167	0.1376	0.1501	7.268	6.663	2.950	19.657	7
8	1.104	0.9054	0.1196	0.1321	8.359	7.568	3.435	25.995	8
9	1.118	0.8942	0.1057	0.1182	9.463	8.462	3.917	33.149	9
10	1.132	0.8832	0.0945	0.1070	10.582	9.346	4.398	41.097	10
11	1.146	0.8723	0.0854	0.0979	11.714	10.218	4.876	49.820	11
12	1.161	0.8615	0.0778	0.0903	12.860	11.079	5.352	59.297	12
13	1.175	0.8509	0.0713	0.0838	14.021	11.930	5.826	69.507	13
14	1.190	0.8404	0.0658	0.0783	15.196	12.771	6.298	80.432	14
15	1.205	0.8300	0.0610	0.0735	16.386	13.601	6.768	92.052	15
16	1.220	0.8197	0.0568	0.0693	17.591	14.420	7.236	104.348	16
17	1.235	0.8096	0.0532	0.0657	18.811	15.230	7.702	117.302	17
18	1.251	0.7996	0.0499	0.0624	20.046	16.030	8.166	130.896	18
19	1.266	0.7898	0.0470	0.0595	21.297	16.819	8.628	145.111	19
20	1.282	0.7800	0.0443	0.0568	22.563	17.599	9.087	159.932	20
21	1.298	0.7704	0.0419	0.0544	23.845	18.370	9.545	175.339	21
22	1.314	0.7609	0.0398	0.0523	25.143	19.131	10.001	191.317	22
23	1.331	0.7515	0.0378	0.0503	26.457	19.882	10.454	207.850	23
24	1.347	0.7422	0.0360	0.0485	27.788	20.624	10.906	224.920	24
25	1.364	0.7330	0.0343	0.0468	29.135	21.357	11.355	242.513	25
26	1.381	0.7240	0.0328	0.0453	30.500	22.081	11.802	260.613	26
27	1.399	0.7150	0.0314	0.0439	31.881	22.796	12.248	279.204	27
28	1.416	0.7062	0.0300	0.0425	33.279	23.503	12.691	298.272	28
29	1.434	0.6975	0.0288	0.0413	34.695	24.200	13.132	317.802	29
30	1.452	0.6889	0.0277	0.0402	36.129	24.889	13.571	337.780	30
31	1.470	0.6804	0.0266	0.0391	37.581	25.569	14.009	358.191	31
32	1.488	0.6720	0.0256	0.0381	39.050	26.241	14.444	379.023	32
33	1.507	0.6637	0.0247	0.0372	40.539	26.905	14.877	400.261	33
34	1.526	0.6555	0.0238	0.0363	42.045	27.560	15.308	421.892	34
35	1.545	0.6474	0.0230	0.0355	43.571	28.208	15.737	443.904	35
36	1.564	0.6394	0.0222	0.0347	45.116	28.847	16.164	466.283	36
37	1.583	0.6315	0.0214	0.0339	46.679	29.479	16.589	489.018	37
38	1.603	0.6237	0.0207	0.0332	48.263	30.103	17.012	512.095	38
39	1.623	0.6160	0.0201	0.0326	49.866	30.719	17.433	535.504	39
40	1.644	0.6084	0.0194	0.0319	51.490	31.327	17.851	559.232	40
41	1.664	0.6009	0.0188	0.0313	53.133	31.928	18.268	583.268	41
42	1.685	0.5935	0.0182	0.0307	54.797	32.521	18.683	607.601	42
43	1.706	0.5862	0.0177	0.0302	56.482	33.107	19.096	632.219	43
44	1.727	0.5789	0.0172	0.0297	58.188	33.686	19.507	657.113	44
45	1.749	0.5718	0.0167	0.0292	59.916	34.258	19.916	682.271	45
46	1.771	0.5647	0.0162	0.0287	61.665	34.823	20.322	707.683	46
47	1.793	0.5577	0.0158	0.0283	63.435	35.381	20.727	733.339	47
48	1.815	0.5509	0.0153	0.0278	65.228	35.931	21.130	759.230	48
49	1.838	0.5441	0.0149	0.0274	67.044	36.476	21.531	785.344	49
50	1.861	0.5373	0.0145	0.0270	68.882	37.013	21.929	811.674	50

1.5%				Compound Interest Factors					1.5%
Period	Single Payment		Uniform Payment Series				Arithmetic Gradient		Period
	Compound Amount Factor	Present Value Factor	Sinking Fund Factor	Capital Recovery Factor	Compound Amount Factor	Present Value Factor	Gradient Uniform Series	Gradient Present Value	
	Find F Given P	Find P Given F	Find A Given F	Find A Given P	Find F Given A	Find P Given A	Find A Given G	Find P Given G	
n	F/P	P/F	A/F	A/P	F/A	P/A	A/G	P/G	n
1	1.015	0.9852	1.0000	1.0150	1.000	0.985	0.000	0.000	1
2	1.030	0.9707	0.4963	0.5113	2.015	1.956	0.496	0.971	2
3	1.046	0.9563	0.3284	0.3434	3.045	2.912	0.990	2.883	3
4	1.061	0.9422	0.2444	0.2594	4.091	3.854	1.481	5.710	4
5	1.077	0.9283	0.1941	0.2091	5.152	4.783	1.970	9.423	5
6	1.093	0.9145	0.1605	0.1755	6.230	5.697	2.457	13.996	6
7	1.110	0.9010	0.1366	0.1516	7.323	6.598	2.940	19.402	7
8	1.126	0.8877	0.1186	0.1336	8.433	7.486	3.422	25.616	8
9	1.143	0.8746	0.1046	0.1196	9.559	8.361	3.901	32.612	9
10	1.161	0.8617	0.0934	0.1084	10.703	9.222	4.377	40.367	10
11	1.178	0.8489	0.0843	0.0993	11.863	10.071	4.851	48.857	11
12	1.196	0.8364	0.0767	0.0917	13.041	10.908	5.323	58.057	12
13	1.214	0.8240	0.0702	0.0852	14.237	11.732	5.792	67.945	13
14	1.232	0.8118	0.0647	0.0797	15.450	12.543	6.258	78.499	14
15	1.250	0.7999	0.0599	0.0749	16.682	13.343	6.722	89.697	15
16	1.269	0.7880	0.0558	0.0708	17.932	14.131	7.184	101.518	16
17	1.288	0.7764	0.0521	0.0671	19.201	14.908	7.643	113.940	17
18	1.307	0.7649	0.0488	0.0638	20.489	15.673	8.100	126.943	18
19	1.327	0.7536	0.0459	0.0609	21.797	16.426	8.554	140.508	19
20	1.347	0.7425	0.0432	0.0582	23.124	17.169	9.006	154.615	20
21	1.367	0.7315	0.0409	0.0559	24.471	17.900	9.455	169.245	21
22	1.388	0.7207	0.0387	0.0537	25.838	18.621	9.902	184.380	22
23	1.408	0.7100	0.0367	0.0517	27.225	19.331	10.346	200.001	23
24	1.430	0.6995	0.0349	0.0499	28.634	20.030	10.788	216.090	24
25	1.451	0.6892	0.0333	0.0483	30.063	20.720	11.228	232.631	25
26	1.473	0.6790	0.0317	0.0467	31.514	21.399	11.665	249.607	26
27	1.495	0.6690	0.0303	0.0453	32.987	22.068	12.099	267.000	27
28	1.517	0.6591	0.0290	0.0440	34.481	22.727	12.531	284.796	28
29	1.540	0.6494	0.0278	0.0428	35.999	23.376	12.961	302.978	29
30	1.563	0.6398	0.0266	0.0416	37.539	24.016	13.388	321.531	30
31	1.587	0.6303	0.0256	0.0406	39.102	24.646	13.813	340.440	31
32	1.610	0.6210	0.0246	0.0396	40.688	25.267	14.236	359.691	32
33	1.634	0.6118	0.0236	0.0386	42.299	25.879	14.656	379.269	33
34	1.659	0.6028	0.0228	0.0378	43.933	26.482	15.073	399.161	34
35	1.684	0.5939	0.0219	0.0369	45.592	27.076	15.488	419.352	35
36	1.709	0.5851	0.0212	0.0362	47.276	27.661	15.901	439.830	36
37	1.735	0.5764	0.0204	0.0354	48.985	28.237	16.311	460.582	37
38	1.761	0.5679	0.0197	0.0347	50.720	28.805	16.719	481.595	38
39	1.787	0.5595	0.0191	0.0341	52.481	29.365	17.125	502.858	39
40	1.814	0.5513	0.0184	0.0334	54.268	29.916	17.528	524.357	40
41	1.841	0.5431	0.0178	0.0328	56.082	30.459	17.928	546.081	41
42	1.869	0.5351	0.0173	0.0323	57.923	30.994	18.327	568.020	42
43	1.897	0.5272	0.0167	0.0317	59.792	31.521	18.723	590.162	43
44	1.925	0.5194	0.0162	0.0312	61.689	32.041	19.116	612.496	44
45	1.954	0.5117	0.0157	0.0307	63.614	32.552	19.507	635.011	45
46	1.984	0.5042	0.0153	0.0303	65.568	33.056	19.896	657.698	46
47	2.013	0.4967	0.0148	0.0298	67.552	33.553	20.283	680.546	47
48	2.043	0.4894	0.0144	0.0294	69.565	34.043	20.667	703.546	48
49	2.074	0.4821	0.0140	0.0290	71.609	34.525	21.048	726.688	49
50	2.105	0.4750	0.0136	0.0286	73.683	35.000	21.428	749.964	50

1.75%			Compound Interest Factors						1.75%
Period	Single Payment		Uniform Payment Series				Arithmetic Gradient		Period
	Compound Amount Factor	Present Value Factor	Sinking Fund Factor	Capital Recovery Factor	Compound Amount Factor	Present Value Factor	Gradient Uniform Series	Gradient Present Value	
	Find F Given P	Find P Given F	Find A Given F	Find A Given P	Find F Given A	Find P Given A	Find A Given G	Find P Given G	
n	F/P	P/F	A/F	A/P	F/A	P/A	A/G	P/G	n
1	1.018	0.9828	1.0000	1.0175	1.000	0.983	0.000	0.000	1
2	1.035	0.9659	0.4957	0.5132	2.018	1.949	0.496	0.966	2
3	1.053	0.9493	0.3276	0.3451	3.053	2.898	0.988	2.864	3
4	1.072	0.9330	0.2435	0.2610	4.106	3.831	1.478	5.663	4
5	1.091	0.9169	0.1931	0.2106	5.178	4.748	1.965	9.331	5
6	1.110	0.9011	0.1595	0.1770	6.269	5.649	2.449	13.837	6
7	1.129	0.8856	0.1355	0.1530	7.378	6.535	2.931	19.151	7
8	1.149	0.8704	0.1175	0.1350	8.508	7.405	3.409	25.243	8
9	1.169	0.8554	0.1036	0.1211	9.656	8.260	3.884	32.087	9
10	1.189	0.8407	0.0924	0.1099	10.825	9.101	4.357	39.654	10
11	1.210	0.8263	0.0832	0.1007	12.015	9.927	4.827	47.916	11
12	1.231	0.8121	0.0756	0.0931	13.225	10.740	5.293	56.849	12
13	1.253	0.7981	0.0692	0.0867	14.457	11.538	5.757	66.426	13
14	1.275	0.7844	0.0637	0.0812	15.710	12.322	6.218	76.623	14
15	1.297	0.7709	0.0589	0.0764	16.984	13.093	6.677	87.415	15
16	1.320	0.7576	0.0547	0.0722	18.282	13.850	7.132	98.779	16
17	1.343	0.7446	0.0510	0.0685	19.602	14.595	7.584	110.693	17
18	1.367	0.7318	0.0477	0.0652	20.945	15.327	8.034	123.133	18
19	1.390	0.7192	0.0448	0.0623	22.311	16.046	8.480	136.078	19
20	1.415	0.7068	0.0422	0.0597	23.702	16.753	8.924	149.508	20
21	1.440	0.6947	0.0398	0.0573	25.116	17.448	9.365	163.401	21
22	1.465	0.6827	0.0377	0.0552	26.556	18.130	9.803	177.738	22
23	1.490	0.6710	0.0357	0.0532	28.021	18.801	10.239	192.500	23
24	1.516	0.6594	0.0339	0.0514	29.511	19.461	10.671	207.667	24
25	1.543	0.6481	0.0322	0.0497	31.027	20.109	11.101	223.221	25
26	1.570	0.6369	0.0307	0.0482	32.570	20.746	11.527	239.145	26
27	1.597	0.6260	0.0293	0.0468	34.140	21.372	11.951	255.421	27
28	1.625	0.6152	0.0280	0.0455	35.738	21.987	12.372	272.032	28
29	1.654	0.6046	0.0268	0.0443	37.363	22.592	12.791	288.962	29
30	1.683	0.5942	0.0256	0.0431	39.017	23.186	13.206	306.195	30
31	1.712	0.5840	0.0246	0.0421	40.700	23.770	13.619	323.716	31
32	1.742	0.5740	0.0236	0.0411	42.412	24.344	14.029	341.510	32
33	1.773	0.5641	0.0226	0.0401	44.154	24.908	14.436	359.561	33
34	1.804	0.5544	0.0218	0.0393	45.927	25.462	14.840	377.857	34
35	1.835	0.5449	0.0210	0.0385	47.731	26.007	15.241	396.382	35
36	1.867	0.5355	0.0202	0.0377	49.566	26.543	15.640	415.125	36
37	1.900	0.5263	0.0194	0.0369	51.434	27.069	16.036	434.071	37
38	1.933	0.5172	0.0187	0.0362	53.334	27.586	16.429	453.209	38
39	1.967	0.5083	0.0181	0.0356	55.267	28.095	16.819	472.526	39
40	2.002	0.4996	0.0175	0.0350	57.234	28.594	17.207	492.011	40
41	2.037	0.4910	0.0169	0.0344	59.236	29.085	17.591	511.651	41
42	2.072	0.4826	0.0163	0.0338	61.272	29.568	17.973	531.436	42
43	2.109	0.4743	0.0158	0.0333	63.345	30.042	18.353	551.355	43
44	2.145	0.4661	0.0153	0.0328	65.453	30.508	18.729	571.398	44
45	2.183	0.4581	0.0148	0.0323	67.599	30.966	19.103	591.554	45
46	2.221	0.4502	0.0143	0.0318	69.782	31.416	19.474	611.813	46
47	2.260	0.4425	0.0139	0.0314	72.003	31.859	19.843	632.167	47
48	2.300	0.4349	0.0135	0.0310	74.263	32.294	20.208	652.605	48
49	2.340	0.4274	0.0131	0.0306	76.562	32.721	20.571	673.120	49
50	2.381	0.4200	0.0127	0.0302	78.902	33.141	20.932	693.701	50

2%	Compound Interest Factors								2%
Period	Single Payment		Uniform Payment Series				Arithmetic Gradient		Period
	Compound Amount Factor	Present Value Factor	Sinking Fund Factor	Capital Recovery Factor	Compound Amount Factor	Present Value Factor	Gradient Uniform Series	Gradient Present Value	
	Find F Given P	Find P Given F	Find A Given F	Find A Given P	Find F Given A	Find P Given A	Find A Given G	Find P Given G	
n	F/P	P/F	A/F	A/P	F/A	P/A	A/G	P/G	n
1	1.020	0.9804	1.0000	1.0200	1.000	0.980	0.000	0.000	1
2	1.040	0.9612	0.4950	0.5150	2.020	1.942	0.495	0.961	2
3	1.061	0.9423	0.3268	0.3468	3.060	2.884	0.987	2.846	3
4	1.082	0.9238	0.2426	0.2626	4.122	3.808	1.475	5.617	4
5	1.104	0.9057	0.1922	0.2122	5.204	4.713	1.960	9.240	5
6	1.126	0.8880	0.1585	0.1785	6.308	5.601	2.442	13.680	6
7	1.149	0.8706	0.1345	0.1545	7.434	6.472	2.921	18.903	7
8	1.172	0.8535	0.1165	0.1365	8.583	7.325	3.396	24.878	8
9	1.195	0.8368	0.1025	0.1225	9.755	8.162	3.868	31.572	9
10	1.219	0.8203	0.0913	0.1113	10.950	8.983	4.337	38.955	10
11	1.243	0.8043	0.0822	0.1022	12.169	9.787	4.802	46.998	11
12	1.268	0.7885	0.0746	0.0946	13.412	10.575	5.264	55.671	12
13	1.294	0.7730	0.0681	0.0881	14.680	11.348	5.723	64.948	13
14	1.319	0.7579	0.0626	0.0826	15.974	12.106	6.179	74.800	14
15	1.346	0.7430	0.0578	0.0778	17.293	12.849	6.631	85.202	15
16	1.373	0.7284	0.0537	0.0737	18.639	13.578	7.080	96.129	16
17	1.400	0.7142	0.0500	0.0700	20.012	14.292	7.526	107.555	17
18	1.428	0.7002	0.0467	0.0667	21.412	14.992	7.968	119.458	18
19	1.457	0.6864	0.0438	0.0638	22.841	15.678	8.407	131.814	19
20	1.486	0.6730	0.0412	0.0612	24.297	16.351	8.843	144.600	20
21	1.516	0.6598	0.0388	0.0588	25.783	17.011	9.276	157.796	21
22	1.546	0.6468	0.0366	0.0566	27.299	17.658	9.705	171.379	22
23	1.577	0.6342	0.0347	0.0547	28.845	18.292	10.132	185.331	23
24	1.608	0.6217	0.0329	0.0529	30.422	18.914	10.555	199.630	24
25	1.641	0.6095	0.0312	0.0512	32.030	19.523	10.974	214.259	25
26	1.673	0.5976	0.0297	0.0497	33.671	20.121	11.391	229.199	26
27	1.707	0.5859	0.0283	0.0483	35.344	20.707	11.804	244.431	27
28	1.741	0.5744	0.0270	0.0470	37.051	21.281	12.214	259.939	28
29	1.776	0.5631	0.0258	0.0458	38.792	21.844	12.621	275.706	29
30	1.811	0.5521	0.0246	0.0446	40.568	22.396	13.025	291.716	30
31	1.848	0.5412	0.0236	0.0436	42.379	22.938	13.426	307.954	31
32	1.885	0.5306	0.0226	0.0426	44.227	23.468	13.823	324.403	32
33	1.922	0.5202	0.0217	0.0417	46.112	23.989	14.217	341.051	33
34	1.961	0.5100	0.0208	0.0408	48.034	24.499	14.608	357.882	34
35	2.000	0.5000	0.0200	0.0400	49.994	24.999	14.996	374.883	35
36	2.040	0.4902	0.0192	0.0392	51.994	25.489	15.381	392.040	36
37	2.081	0.4806	0.0185	0.0385	54.034	25.969	15.762	409.342	37
38	2.122	0.4712	0.0178	0.0378	56.115	26.441	16.141	426.776	38
39	2.165	0.4619	0.0172	0.0372	58.237	26.903	16.516	444.330	39
40	2.208	0.4529	0.0166	0.0366	60.402	27.355	16.889	461.993	40
41	2.252	0.4440	0.0160	0.0360	62.610	27.799	17.258	479.754	41
42	2.297	0.4353	0.0154	0.0354	64.862	28.235	17.624	497.601	42
43	2.343	0.4268	0.0149	0.0349	67.159	28.662	17.987	515.525	43
44	2.390	0.4184	0.0144	0.0344	69.503	29.080	18.347	533.517	44
45	2.438	0.4102	0.0139	0.0339	71.893	29.490	18.703	551.565	45
46	2.487	0.4022	0.0135	0.0335	74.331	29.892	19.057	569.662	46
47	2.536	0.3943	0.0130	0.0330	76.817	30.287	19.408	587.798	47
48	2.587	0.3865	0.0126	0.0326	79.354	30.673	19.756	605.966	48
49	2.639	0.3790	0.0122	0.0322	81.941	31.052	20.100	624.156	49
50	2.692	0.3715	0.0118	0.0318	84.579	31.424	20.442	642.361	50

2.5%				Compound Interest Factors					2.5%
Period	Single Payment		Uniform Payment Series				Arithmetic Gradient		Period
	Compound Amount Factor	Present Value Factor	Sinking Fund Factor	Capital Recovery Factor	Compound Amount Factor	Present Value Factor	Gradient Uniform Series	Gradient Present Value	
	Find F Given P	Find P Given F	Find A Given F	Find A Given P	Find F Given A	Find P Given A	Find A Given G	Find P Given G	
n	F/P	P/F	A/F	A/P	F/A	P/A	A/G	P/G	n
1	1.025	0.9756	1.0000	1.0250	1.000	0.976	0.000	0.000	1
2	1.051	0.9518	0.4938	0.5188	2.025	1.927	0.494	0.952	2
3	1.077	0.9286	0.3251	0.3501	3.076	2.856	0.984	2.809	3
4	1.104	0.9060	0.2408	0.2658	4.153	3.762	1.469	5.527	4
5	1.131	0.8839	0.1902	0.2152	5.256	4.646	1.951	9.062	5
6	1.160	0.8623	0.1565	0.1815	6.388	5.508	2.428	13.374	6
7	1.189	0.8413	0.1325	0.1575	7.547	6.349	2.901	18.421	7
8	1.218	0.8207	0.1145	0.1395	8.736	7.170	3.370	24.167	8
9	1.249	0.8007	0.1005	0.1255	9.955	7.971	3.836	30.572	9
10	1.280	0.7812	0.0893	0.1143	11.203	8.752	4.296	37.603	10
11	1.312	0.7621	0.0801	0.1051	12.483	9.514	4.753	45.225	11
12	1.345	0.7436	0.0725	0.0975	13.796	10.258	5.206	53.404	12
13	1.379	0.7254	0.0660	0.0910	15.140	10.983	5.655	62.109	13
14	1.413	0.7077	0.0605	0.0855	16.519	11.691	6.100	71.309	14
15	1.448	0.6905	0.0558	0.0808	17.932	12.381	6.540	80.976	15
16	1.485	0.6736	0.0516	0.0766	19.380	13.055	6.977	91.080	16
17	1.522	0.6572	0.0479	0.0729	20.865	13.712	7.409	101.595	17
18	1.560	0.6412	0.0447	0.0697	22.386	14.353	7.838	112.495	18
19	1.599	0.6255	0.0418	0.0668	23.946	14.979	8.262	123.755	19
20	1.639	0.6103	0.0391	0.0641	25.545	15.589	8.682	135.350	20
21	1.680	0.5954	0.0368	0.0618	27.183	16.185	9.099	147.257	21
22	1.722	0.5809	0.0346	0.0596	28.863	16.765	9.511	159.456	22
23	1.765	0.5667	0.0327	0.0577	30.584	17.332	9.919	171.923	23
24	1.809	0.5529	0.0309	0.0559	32.349	17.885	10.324	184.639	24
25	1.854	0.5394	0.0293	0.0543	34.158	18.424	10.724	197.584	25
26	1.900	0.5262	0.0278	0.0528	36.012	18.951	11.121	210.740	26
27	1.948	0.5134	0.0264	0.0514	37.912	19.464	11.513	224.089	27
28	1.996	0.5009	0.0251	0.0501	39.860	19.965	11.902	237.612	28
29	2.046	0.4887	0.0239	0.0489	41.856	20.454	12.286	251.295	29
30	2.098	0.4767	0.0228	0.0478	43.903	20.930	12.667	265.120	30
31	2.150	0.4651	0.0217	0.0467	46.000	21.395	13.044	279.074	31
32	2.204	0.4538	0.0208	0.0458	48.150	21.849	13.417	293.141	32
33	2.259	0.4427	0.0199	0.0449	50.354	22.292	13.786	307.307	33
34	2.315	0.4319	0.0190	0.0440	52.613	22.724	14.151	321.560	34
35	2.373	0.4214	0.0182	0.0432	54.928	23.145	14.512	335.887	35
36	2.433	0.4111	0.0175	0.0425	57.301	23.556	14.870	350.275	36
37	2.493	0.4011	0.0167	0.0417	59.734	23.957	15.223	364.713	37
38	2.556	0.3913	0.0161	0.0411	62.227	24.349	15.573	379.191	38
39	2.620	0.3817	0.0154	0.0404	64.783	24.730	15.920	393.697	39
40	2.685	0.3724	0.0148	0.0398	67.403	25.103	16.262	408.222	40
41	2.752	0.3633	0.0143	0.0393	70.088	25.466	16.601	422.756	41
42	2.821	0.3545	0.0137	0.0387	72.840	25.821	16.936	437.290	42
43	2.892	0.3458	0.0132	0.0382	75.661	26.166	17.267	451.815	43
44	2.964	0.3374	0.0127	0.0377	78.552	26.504	17.595	466.323	44
45	3.038	0.3292	0.0123	0.0373	81.516	26.833	17.918	480.807	45
46	3.114	0.3211	0.0118	0.0368	84.554	27.154	18.239	495.259	46
47	3.192	0.3133	0.0114	0.0364	87.668	27.467	18.555	509.671	47
48	3.271	0.3057	0.0110	0.0360	90.860	27.773	18.868	524.038	48
49	3.353	0.2982	0.0106	0.0356	94.131	28.071	19.178	538.352	49
50	3.437	0.2909	0.0103	0.0353	97.484	28.362	19.484	552.608	50

3%	Compound Interest Factors								3%
Period	Single Payment		Uniform Payment Series				Arithmetic Gradient		Period
	Compound Amount Factor	Present Value Factor	Sinking Fund Factor	Capital Recovery Factor	Compound Amount Factor	Present Value Factor	Gradient Uniform Series	Gradient Present Value	
	Find F Given P	Find P Given F	Find A Given F	Find A Given P	Find F Given A	Find P Given A	Find A Given G	Find P Given G	
n	F/P	P/F	A/F	A/P	F/A	P/A	A/G	P/G	n
1	1.030	0.9709	1.0000	1.0300	1.000	0.971	0.000	0.000	1
2	1.061	0.9426	0.4926	0.5226	2.030	1.913	0.493	0.943	2
3	1.093	0.9151	0.3235	0.3535	3.091	2.829	0.980	2.773	3
4	1.126	0.8885	0.2390	0.2690	4.184	3.717	1.463	5.438	4
5	1.159	0.8626	0.1884	0.2184	5.309	4.580	1.941	8.889	5
6	1.194	0.8375	0.1546	0.1846	6.468	5.417	2.414	13.076	6
7	1.230	0.8131	0.1305	0.1605	7.662	6.230	2.882	17.955	7
8	1.267	0.7894	0.1125	0.1425	8.892	7.020	3.345	23.481	8
9	1.305	0.7664	0.0984	0.1284	10.159	7.786	3.803	29.612	9
10	1.344	0.7441	0.0872	0.1172	11.464	8.530	4.256	36.309	10
11	1.384	0.7224	0.0781	0.1081	12.808	9.253	4.705	43.533	11
12	1.426	0.7014	0.0705	0.1005	14.192	9.954	5.148	51.248	12
13	1.469	0.6810	0.0640	0.0940	15.618	10.635	5.587	59.420	13
14	1.513	0.6611	0.0585	0.0885	17.086	11.296	6.021	68.014	14
15	1.558	0.6419	0.0538	0.0838	18.599	11.938	6.450	77.000	15
16	1.605	0.6232	0.0496	0.0796	20.157	12.561	6.874	86.348	16
17	1.653	0.6050	0.0460	0.0760	21.762	13.166	7.294	96.028	17
18	1.702	0.5874	0.0427	0.0727	23.414	13.754	7.708	106.014	18
19	1.754	0.5703	0.0398	0.0698	25.117	14.324	8.118	116.279	19
20	1.806	0.5537	0.0372	0.0672	26.870	14.877	8.523	126.799	20
21	1.860	0.5375	0.0349	0.0649	28.676	15.415	8.923	137.550	21
22	1.916	0.5219	0.0327	0.0627	30.537	15.937	9.319	148.509	22
23	1.974	0.5067	0.0308	0.0608	32.453	16.444	9.709	159.657	23
24	2.033	0.4919	0.0290	0.0590	34.426	16.936	10.095	170.971	24
25	2.094	0.4776	0.0274	0.0574	36.459	17.413	10.477	182.434	25
26	2.157	0.4637	0.0259	0.0559	38.553	17.877	10.853	194.026	26
27	2.221	0.4502	0.0246	0.0546	40.710	18.327	11.226	205.731	27
28	2.288	0.4371	0.0233	0.0533	42.931	18.764	11.593	217.532	28
29	2.357	0.4243	0.0221	0.0521	45.219	19.188	11.956	229.414	29
30	2.427	0.4120	0.0210	0.0510	47.575	19.600	12.314	241.361	30
31	2.500	0.4000	0.0200	0.0500	50.003	20.000	12.668	253.361	31
32	2.575	0.3883	0.0190	0.0490	52.503	20.389	13.017	265.399	32
33	2.652	0.3770	0.0182	0.0482	55.078	20.766	13.362	277.464	33
34	2.732	0.3660	0.0173	0.0473	57.730	21.132	13.702	289.544	34
35	2.814	0.3554	0.0165	0.0465	60.462	21.487	14.037	301.627	35
36	2.898	0.3450	0.0158	0.0458	63.276	21.832	14.369	313.703	36
37	2.985	0.3350	0.0151	0.0451	66.174	22.167	14.696	325.762	37
38	3.075	0.3252	0.0145	0.0445	69.159	22.492	15.018	337.796	38
39	3.167	0.3158	0.0138	0.0438	72.234	22.808	15.336	349.794	39
40	3.262	0.3066	0.0133	0.0433	75.401	23.115	15.650	361.750	40
41	3.360	0.2976	0.0127	0.0427	78.663	23.412	15.960	373.655	41
42	3.461	0.2890	0.0122	0.0422	82.023	23.701	16.265	385.502	42
43	3.565	0.2805	0.0117	0.0417	85.484	23.982	16.566	397.285	43
44	3.671	0.2724	0.0112	0.0412	89.048	24.254	16.863	408.997	44
45	3.782	0.2644	0.0108	0.0408	92.720	24.519	17.156	420.632	45
46	3.895	0.2567	0.0104	0.0404	96.501	24.775	17.444	432.186	46
47	4.012	0.2493	0.0100	0.0400	100.397	25.025	17.729	443.652	47
48	4.132	0.2420	0.0096	0.0396	104.408	25.267	18.009	455.025	48
49	4.256	0.2350	0.0092	0.0392	108.541	25.502	18.285	466.303	49
50	4.384	0.2281	0.0089	0.0389	112.797	25.730	18.558	477.480	50

3.5%				Compound Interest Factors					3.5%
Period	Single Payment		Uniform Payment Series				Arithmetic Gradient		Period
	Compound Amount Factor	Present Value Factor	Sinking Fund Factor	Capital Recovery Factor	Compound Amount Factor	Present Value Factor	Gradient Uniform Series	Gradient Present Value	
	Find F Given P	Find P Given F	Find A Given F	Find A Given P	Find F Given A	Find P Given A	Find A Given G	Find P Given G	
n	F/P	P/F	A/F	A/P	F/A	P/A	A/G	P/G	n
1	1.035	0.9662	1.0000	1.0350	1.000	0.966	0.000	0.000	1
2	1.071	0.9335	0.4914	0.5264	2.035	1.900	0.491	0.934	2
3	1.109	0.9019	0.3219	0.3569	3.106	2.802	0.977	2.737	3
4	1.148	0.8714	0.2373	0.2723	4.215	3.673	1.457	5.352	4
5	1.188	0.8420	0.1865	0.2215	5.362	4.515	1.931	8.720	5
6	1.229	0.8135	0.1527	0.1877	6.550	5.329	2.400	12.787	6
7	1.272	0.7860	0.1285	0.1635	7.779	6.115	2.863	17.503	7
8	1.317	0.7594	0.1105	0.1455	9.052	6.874	3.320	22.819	8
9	1.363	0.7337	0.0964	0.1314	10.368	7.608	3.771	28.689	9
10	1.411	0.7089	0.0852	0.1202	11.731	8.317	4.217	35.069	10
11	1.460	0.6849	0.0761	0.1111	13.142	9.002	4.657	41.919	11
12	1.511	0.6618	0.0685	0.1035	14.602	9.663	5.091	49.198	12
13	1.564	0.6394	0.0621	0.0971	16.113	10.303	5.520	56.871	13
14	1.619	0.6178	0.0566	0.0916	17.677	10.921	5.943	64.902	14
15	1.675	0.5969	0.0518	0.0868	19.296	11.517	6.361	73.259	15
16	1.734	0.5767	0.0477	0.0827	20.971	12.094	6.773	81.909	16
17	1.795	0.5572	0.0440	0.0790	22.705	12.651	7.179	90.824	17
18	1.857	0.5384	0.0408	0.0758	24.500	13.190	7.580	99.977	18
19	1.923	0.5202	0.0379	0.0729	26.357	13.710	7.975	109.339	19
20	1.990	0.5026	0.0354	0.0704	28.280	14.212	8.365	118.888	20
21	2.059	0.4856	0.0330	0.0680	30.269	14.698	8.749	128.600	21
22	2.132	0.4692	0.0309	0.0659	32.329	15.167	9.128	138.452	22
23	2.206	0.4533	0.0290	0.0640	34.460	15.620	9.502	148.424	23
24	2.283	0.4380	0.0273	0.0623	36.667	16.058	9.870	158.497	24
25	2.363	0.4231	0.0257	0.0607	38.950	16.482	10.233	168.653	25
26	2.446	0.4088	0.0242	0.0592	41.313	16.890	10.590	178.874	26
27	2.532	0.3950	0.0229	0.0579	43.759	17.285	10.942	189.144	27
28	2.620	0.3817	0.0216	0.0566	46.291	17.667	11.289	199.448	28
29	2.712	0.3687	0.0204	0.0554	48.911	18.036	11.631	209.773	29
30	2.807	0.3563	0.0194	0.0544	51.623	18.392	11.967	220.106	30
31	2.905	0.3442	0.0184	0.0534	54.429	18.736	12.299	230.432	31
32	3.007	0.3326	0.0174	0.0524	57.335	19.069	12.625	240.743	32
33	3.112	0.3213	0.0166	0.0516	60.341	19.390	12.946	251.026	33
34	3.221	0.3105	0.0158	0.0508	63.453	19.701	13.262	261.271	34
35	3.334	0.3000	0.0150	0.0500	66.674	20.001	13.573	271.471	35
36	3.450	0.2898	0.0143	0.0493	70.008	20.290	13.879	281.615	36
37	3.571	0.2800	0.0136	0.0486	73.458	20.571	14.180	291.696	37
38	3.696	0.2706	0.0130	0.0480	77.029	20.841	14.477	301.707	38
39	3.825	0.2614	0.0124	0.0474	80.725	21.102	14.768	311.640	39
40	3.959	0.2526	0.0118	0.0468	84.550	21.355	15.055	321.491	40
41	4.098	0.2440	0.0113	0.0463	88.510	21.599	15.336	331.252	41
42	4.241	0.2358	0.0108	0.0458	92.607	21.835	15.613	340.919	42
43	4.390	0.2278	0.0103	0.0453	96.849	22.063	15.886	350.487	43
44	4.543	0.2201	0.0099	0.0449	101.238	22.283	16.154	359.951	44
45	4.702	0.2127	0.0095	0.0445	105.782	22.495	16.417	369.308	45
46	4.867	0.2055	0.0091	0.0441	110.484	22.701	16.676	378.554	46
47	5.037	0.1985	0.0087	0.0437	115.351	22.899	16.930	387.686	47
48	5.214	0.1918	0.0083	0.0433	120.388	23.091	17.180	396.701	48
49	5.396	0.1853	0.0080	0.0430	125.602	23.277	17.425	405.596	49
50	5.585	0.1791	0.0076	0.0426	130.998	23.456	17.666	414.370	50

4%				Compound Interest Factors					4%
Period	Single Payment		Uniform Payment Series				Arithmetic Gradient		Period
	Compound Amount Factor	Present Value Factor	Sinking Fund Factor	Capital Recovery Factor	Compound Amount Factor	Present Value Factor	Gradient Uniform Series	Gradient Present Value	
	Find F Given P	Find P Given F	Find A Given F	Find A Given P	Find F Given A	Find P Given A	Find A Given G	Find P Given G	
n	F/P	P/F	A/F	A/P	F/A	P/A	A/G	P/G	n
1	1.040	0.9615	1.0000	1.0400	1.000	0.962	0.000	0.000	1
2	1.082	0.9246	0.4902	0.5302	2.040	1.886	0.490	0.925	2
3	1.125	0.8890	0.3203	0.3603	3.122	2.775	0.974	2.703	3
4	1.170	0.8548	0.2355	0.2755	4.246	3.630	1.451	5.267	4
5	1.217	0.8219	0.1846	0.2246	5.416	4.452	1.922	8.555	5
6	1.265	0.7903	0.1508	0.1908	6.633	5.242	2.386	12.506	6
7	1.316	0.7599	0.1266	0.1666	7.898	6.002	2.843	17.066	7
8	1.369	0.7307	0.1085	0.1485	9.214	6.733	3.294	22.181	8
9	1.423	0.7026	0.0945	0.1345	10.583	7.435	3.739	27.801	9
10	1.480	0.6756	0.0833	0.1233	12.006	8.111	4.177	33.881	10
11	1.539	0.6496	0.0741	0.1141	13.486	8.760	4.609	40.377	11
12	1.601	0.6246	0.0666	0.1066	15.026	9.385	5.034	47.248	12
13	1.665	0.6006	0.0601	0.1001	16.627	9.986	5.453	54.455	13
14	1.732	0.5775	0.0547	0.0947	18.292	10.563	5.866	61.962	14
15	1.801	0.5553	0.0499	0.0899	20.024	11.118	6.272	69.735	15
16	1.873	0.5339	0.0458	0.0858	21.825	11.652	6.672	77.744	16
17	1.948	0.5134	0.0422	0.0822	23.698	12.166	7.066	85.958	17
18	2.026	0.4936	0.0390	0.0790	25.645	12.659	7.453	94.350	18
19	2.107	0.4746	0.0361	0.0761	27.671	13.134	7.834	102.893	19
20	2.191	0.4564	0.0336	0.0736	29.778	13.590	8.209	111.565	20
21	2.279	0.4388	0.0313	0.0713	31.969	14.029	8.578	120.341	21
22	2.370	0.4220	0.0292	0.0692	34.248	14.451	8.941	129.202	22
23	2.465	0.4057	0.0273	0.0673	36.618	14.857	9.297	138.128	23
24	2.563	0.3901	0.0256	0.0656	39.083	15.247	9.648	147.101	24
25	2.666	0.3751	0.0240	0.0640	41.646	15.622	9.993	156.104	25
26	2.772	0.3607	0.0226	0.0626	44.312	15.983	10.331	165.121	26
27	2.883	0.3468	0.0212	0.0612	47.084	16.330	10.664	174.138	27
28	2.999	0.3335	0.0200	0.0600	49.968	16.663	10.991	183.142	28
29	3.119	0.3207	0.0189	0.0589	52.966	16.984	11.312	192.121	29
30	3.243	0.3083	0.0178	0.0578	56.085	17.292	11.627	201.062	30
31	3.373	0.2965	0.0169	0.0569	59.328	17.588	11.937	209.956	31
32	3.508	0.2851	0.0159	0.0559	62.701	17.874	12.241	218.792	32
33	3.648	0.2741	0.0151	0.0551	66.210	18.148	12.540	227.563	33
34	3.794	0.2636	0.0143	0.0543	69.858	18.411	12.832	236.261	34
35	3.946	0.2534	0.0136	0.0536	73.652	18.665	13.120	244.877	35
36	4.104	0.2437	0.0129	0.0529	77.598	18.908	13.402	253.405	36
37	4.268	0.2343	0.0122	0.0522	81.702	19.143	13.678	261.840	37
38	4.439	0.2253	0.0116	0.0516	85.970	19.368	13.950	270.175	38
39	4.616	0.2166	0.0111	0.0511	90.409	19.584	14.216	278.407	39
40	4.801	0.2083	0.0105	0.0505	95.026	19.793	14.477	286.530	40
41	4.993	0.2003	0.0100	0.0500	99.827	19.993	14.732	294.541	41
42	5.193	0.1926	0.0095	0.0495	104.820	20.186	14.983	302.437	42
43	5.400	0.1852	0.0091	0.0491	110.012	20.371	15.228	310.214	43
44	5.617	0.1780	0.0087	0.0487	115.413	20.549	15.469	317.870	44
45	5.841	0.1712	0.0083	0.0483	121.029	20.720	15.705	325.403	45
46	6.075	0.1646	0.0079	0.0479	126.871	20.885	15.936	332.810	46
47	6.318	0.1583	0.0075	0.0475	132.945	21.043	16.162	340.091	47
48	6.571	0.1522	0.0072	0.0472	139.263	21.195	16.383	347.245	48
49	6.833	0.1463	0.0069	0.0469	145.834	21.341	16.600	354.269	49
50	7.107	0.1407	0.0066	0.0466	152.667	21.482	16.812	361.164	50

4.5%				Compound Interest Factors					4.5%
Period	Single Payment		Uniform Payment Series				Arithmetic Gradient		Period
	Compound Amount Factor	Present Value Factor	Sinking Fund Factor	Capital Recovery Factor	Compound Amount Factor	Present Value Factor	Gradient Uniform Series	Gradient Present Value	
	Find F Given P	Find P Given F	Find A Given F	Find A Given P	Find F Given A	Find P Given A	Find A Given G	Find P Given G	
n	F/P	P/F	A/F	A/P	F/A	P/A	A/G	P/G	n
1	1.045	0.9569	1.0000	1.0450	1.000	0.957	0.000	0.000	1
2	1.092	0.9157	0.4890	0.5340	2.045	1.873	0.489	0.916	2
3	1.141	0.8763	0.3188	0.3638	3.137	2.749	0.971	2.668	3
4	1.193	0.8386	0.2337	0.2787	4.278	3.588	1.445	5.184	4
5	1.246	0.8025	0.1828	0.2278	5.471	4.390	1.912	8.394	5
6	1.302	0.7679	0.1489	0.1939	6.717	5.158	2.372	12.233	6
7	1.361	0.7348	0.1247	0.1697	8.019	5.893	2.824	16.642	7
8	1.422	0.7032	0.1066	0.1516	9.380	6.596	3.269	21.565	8
9	1.486	0.6729	0.0926	0.1376	10.802	7.269	3.707	26.948	9
10	1.553	0.6439	0.0814	0.1264	12.288	7.913	4.138	32.743	10
11	1.623	0.6162	0.0722	0.1172	13.841	8.529	4.562	38.905	11
12	1.696	0.5897	0.0647	0.1097	15.464	9.119	4.978	45.391	12
13	1.772	0.5643	0.0583	0.1033	17.160	9.683	5.387	52.163	13
14	1.852	0.5400	0.0528	0.0978	18.932	10.223	5.789	59.182	14
15	1.935	0.5167	0.0481	0.0931	20.784	10.740	6.184	66.416	15
16	2.022	0.4945	0.0440	0.0890	22.719	11.234	6.572	73.833	16
17	2.113	0.4732	0.0404	0.0854	24.742	11.707	6.953	81.404	17
18	2.208	0.4528	0.0372	0.0822	26.855	12.160	7.327	89.102	18
19	2.308	0.4333	0.0344	0.0794	29.064	12.593	7.695	96.901	19
20	2.412	0.4146	0.0319	0.0769	31.371	13.008	8.055	104.780	20
21	2.520	0.3968	0.0296	0.0746	33.783	13.405	8.409	112.715	21
22	2.634	0.3797	0.0275	0.0725	36.303	13.784	8.755	120.689	22
23	2.752	0.3634	0.0257	0.0707	38.937	14.148	9.096	128.683	23
24	2.876	0.3477	0.0240	0.0690	41.689	14.495	9.429	136.680	24
25	3.005	0.3327	0.0224	0.0674	44.565	14.828	9.756	144.665	25
26	3.141	0.3184	0.0210	0.0660	47.571	15.147	10.077	152.625	26
27	3.282	0.3047	0.0197	0.0647	50.711	15.451	10.391	160.547	27
28	3.430	0.2916	0.0185	0.0635	53.993	15.743	10.698	168.420	28
29	3.584	0.2790	0.0174	0.0624	57.423	16.022	10.999	176.232	29
30	3.745	0.2670	0.0164	0.0614	61.007	16.289	11.295	183.975	30
31	3.914	0.2555	0.0154	0.0604	64.752	16.544	11.583	191.640	31
32	4.090	0.2445	0.0146	0.0596	68.666	16.789	11.866	199.220	32
33	4.274	0.2340	0.0137	0.0587	72.756	17.023	12.143	206.707	33
34	4.466	0.2239	0.0130	0.0580	77.030	17.247	12.414	214.096	34
35	4.667	0.2143	0.0123	0.0573	81.497	17.461	12.679	221.380	35
36	4.877	0.2050	0.0116	0.0566	86.164	17.666	12.938	228.556	36
37	5.097	0.1962	0.0110	0.0560	91.041	17.862	13.191	235.619	37
38	5.326	0.1878	0.0104	0.0554	96.138	18.050	13.439	242.566	38
39	5.566	0.1797	0.0099	0.0549	101.464	18.230	13.681	249.393	39
40	5.816	0.1719	0.0093	0.0543	107.030	18.402	13.917	256.099	40
41	6.078	0.1645	0.0089	0.0539	112.847	18.566	14.148	262.680	41
42	6.352	0.1574	0.0084	0.0534	118.925	18.724	14.374	269.135	42
43	6.637	0.1507	0.0080	0.0530	125.276	18.874	14.595	275.462	43
44	6.936	0.1442	0.0076	0.0526	131.914	19.018	14.810	281.662	44
45	7.248	0.1380	0.0072	0.0522	138.850	19.156	15.020	287.732	45
46	7.574	0.1320	0.0068	0.0518	146.098	19.288	15.225	293.673	46
47	7.915	0.1263	0.0065	0.0515	153.673	19.415	15.426	299.485	47
48	8.271	0.1209	0.0062	0.0512	161.588	19.536	15.621	305.167	48
49	8.644	0.1157	0.0059	0.0509	169.859	19.651	15.812	310.720	49
50	9.033	0.1107	0.0056	0.0506	178.503	19.762	15.998	316.145	50

5%	Compound Interest Factors								5%
Period	Single Payment		Uniform Payment Series				Arithmetic Gradient		Period
	Compound Amount Factor	Present Value Factor	Sinking Fund Factor	Capital Recovery Factor	Compound Amount Factor	Present Value Factor	Gradient Uniform Series	Gradient Present Value	
	Find F Given P	Find P Given F	Find A Given F	Find A Given P	Find F Given A	Find P Given A	Find A Given G	Find P Given G	
n	F/P	P/F	A/F	A/P	F/A	P/A	A/G	P/G	n
1	1.050	0.9524	1.0000	1.0500	1.000	0.952	0.000	0.000	1
2	1.103	0.9070	0.4878	0.5378	2.050	1.859	0.488	0.907	2
3	1.158	0.8638	0.3172	0.3672	3.153	2.723	0.967	2.635	3
4	1.216	0.8227	0.2320	0.2820	4.310	3.546	1.439	5.103	4
5	1.276	0.7835	0.1810	0.2310	5.526	4.329	1.903	8.237	5
6	1.340	0.7462	0.1470	0.1970	6.802	5.076	2.358	11.968	6
7	1.407	0.7107	0.1228	0.1728	8.142	5.786	2.805	16.232	7
8	1.477	0.6768	0.1047	0.1547	9.549	6.463	3.245	20.970	8
9	1.551	0.6446	0.0907	0.1407	11.027	7.108	3.676	26.127	9
10	1.629	0.6139	0.0795	0.1295	12.578	7.722	4.099	31.652	10
11	1.710	0.5847	0.0704	0.1204	14.207	8.306	4.514	37.499	11
12	1.796	0.5568	0.0628	0.1128	15.917	8.863	4.922	43.624	12
13	1.886	0.5303	0.0565	0.1065	17.713	9.394	5.322	49.988	13
14	1.980	0.5051	0.0510	0.1010	19.599	9.899	5.713	56.554	14
15	2.079	0.4810	0.0463	0.0963	21.579	10.380	6.097	63.288	15
16	2.183	0.4581	0.0423	0.0923	23.657	10.838	6.474	70.160	16
17	2.292	0.4363	0.0387	0.0887	25.840	11.274	6.842	77.140	17
18	2.407	0.4155	0.0355	0.0855	28.132	11.690	7.203	84.204	18
19	2.527	0.3957	0.0327	0.0827	30.539	12.085	7.557	91.328	19
20	2.653	0.3769	0.0302	0.0802	33.066	12.462	7.903	98.488	20
21	2.786	0.3589	0.0280	0.0780	35.719	12.821	8.242	105.667	21
22	2.925	0.3418	0.0260	0.0760	38.505	13.163	8.573	112.846	22
23	3.072	0.3256	0.0241	0.0741	41.430	13.489	8.897	120.009	23
24	3.225	0.3101	0.0225	0.0725	44.502	13.799	9.214	127.140	24
25	3.386	0.2953	0.0210	0.0710	47.727	14.094	9.524	134.228	25
26	3.556	0.2812	0.0196	0.0696	51.113	14.375	9.827	141.259	26
27	3.733	0.2678	0.0183	0.0683	54.669	14.643	10.122	148.223	27
28	3.920	0.2551	0.0171	0.0671	58.403	14.898	10.411	155.110	28
29	4.116	0.2429	0.0160	0.0660	62.323	15.141	10.694	161.913	29
30	4.322	0.2314	0.0151	0.0651	66.439	15.372	10.969	168.623	30
31	4.538	0.2204	0.0141	0.0641	70.761	15.593	11.238	175.233	31
32	4.765	0.2099	0.0133	0.0633	75.299	15.803	11.501	181.739	32
33	5.003	0.1999	0.0125	0.0625	80.064	16.003	11.757	188.135	33
34	5.253	0.1904	0.0118	0.0618	85.067	16.193	12.006	194.417	34
35	5.516	0.1813	0.0111	0.0611	90.320	16.374	12.250	200.581	35
36	5.792	0.1727	0.0104	0.0604	95.836	16.547	12.487	206.624	36
37	6.081	0.1644	0.0098	0.0598	101.628	16.711	12.719	212.543	37
38	6.385	0.1566	0.0093	0.0593	107.710	16.868	12.944	218.338	38
39	6.705	0.1491	0.0088	0.0588	114.095	17.017	13.164	224.005	39
40	7.040	0.1420	0.0083	0.0583	120.800	17.159	13.377	229.545	40
41	7.392	0.1353	0.0078	0.0578	127.840	17.294	13.586	234.956	41
42	7.762	0.1288	0.0074	0.0574	135.232	17.423	13.788	240.239	42
43	8.150	0.1227	0.0070	0.0570	142.993	17.546	13.986	245.392	43
44	8.557	0.1169	0.0066	0.0566	151.143	17.663	14.178	250.417	44
45	8.985	0.1113	0.0063	0.0563	159.700	17.774	14.364	255.315	45
46	9.434	0.1060	0.0059	0.0559	168.685	17.880	14.546	260.084	46
47	9.906	0.1009	0.0056	0.0556	178.119	17.981	14.723	264.728	47
48	10.401	0.0961	0.0053	0.0553	188.025	18.077	14.894	269.247	48
49	10.921	0.0916	0.0050	0.0550	198.427	18.169	15.061	273.642	49
50	11.467	0.0872	0.0048	0.0548	209.348	18.256	15.223	277.915	50

6%				Compound Interest Factors					6%
Period	Single Payment		Uniform Payment Series				Arithmetic Gradient		Period
	Compound Amount Factor	Present Value Factor	Sinking Fund Factor	Capital Recovery Factor	Compound Amount Factor	Present Value Factor	Gradient Uniform Series	Gradient Present Value	
	Find F Given P	Find P Given F	Find A Given F	Find A Given P	Find F Given A	Find P Given A	Find A Given G	Find P Given G	
n	F/P	P/F	A/F	A/P	F/A	P/A	A/G	P/G	n
1	1.060	0.9434	1.0000	1.0600	1.000	0.943	0.000	0.000	1
2	1.124	0.8900	0.4854	0.5454	2.060	1.833	0.485	0.890	2
3	1.191	0.8396	0.3141	0.3741	3.184	2.673	0.961	2.569	3
4	1.262	0.7921	0.2286	0.2886	4.375	3.465	1.427	4.946	4
5	1.338	0.7473	0.1774	0.2374	5.637	4.212	1.884	7.935	5
6	1.419	0.7050	0.1434	0.2034	6.975	4.917	2.330	11.459	6
7	1.504	0.6651	0.1191	0.1791	8.394	5.582	2.768	15.450	7
8	1.594	0.6274	0.1010	0.1610	9.897	6.210	3.195	19.842	8
9	1.689	0.5919	0.0870	0.1470	11.491	6.802	3.613	24.577	9
10	1.791	0.5584	0.0759	0.1359	13.181	7.360	4.022	29.602	10
11	1.898	0.5268	0.0668	0.1268	14.972	7.887	4.421	34.870	11
12	2.012	0.4970	0.0593	0.1193	16.870	8.384	4.811	40.337	12
13	2.133	0.4688	0.0530	0.1130	18.882	8.853	5.192	45.963	13
14	2.261	0.4423	0.0476	0.1076	21.015	9.295	5.564	51.713	14
15	2.397	0.4173	0.0430	0.1030	23.276	9.712	5.926	57.555	15
16	2.540	0.3936	0.0390	0.0990	25.673	10.106	6.279	63.459	16
17	2.693	0.3714	0.0354	0.0954	28.213	10.477	6.624	69.401	17
18	2.854	0.3503	0.0324	0.0924	30.906	10.828	6.960	75.357	18
19	3.026	0.3305	0.0296	0.0896	33.760	11.158	7.287	81.306	19
20	3.207	0.3118	0.0272	0.0872	36.786	11.470	7.605	87.230	20
21	3.400	0.2942	0.0250	0.0850	39.993	11.764	7.915	93.114	21
22	3.604	0.2775	0.0230	0.0830	43.392	12.042	8.217	98.941	22
23	3.820	0.2618	0.0213	0.0813	46.996	12.303	8.510	104.701	23
24	4.049	0.2470	0.0197	0.0797	50.816	12.550	8.795	110.381	24
25	4.292	0.2330	0.0182	0.0782	54.865	12.783	9.072	115.973	25
26	4.549	0.2198	0.0169	0.0769	59.156	13.003	9.341	121.468	26
27	4.822	0.2074	0.0157	0.0757	63.706	13.211	9.603	126.860	27
28	5.112	0.1956	0.0146	0.0746	68.528	13.406	9.857	132.142	28
29	5.418	0.1846	0.0136	0.0736	73.640	13.591	10.103	137.310	29
30	5.743	0.1741	0.0126	0.0726	79.058	13.765	10.342	142.359	30
31	6.088	0.1643	0.0118	0.0718	84.802	13.929	10.574	147.286	31
32	6.453	0.1550	0.0110	0.0710	90.890	14.084	10.799	152.090	32
33	6.841	0.1462	0.0103	0.0703	97.343	14.230	11.017	156.768	33
34	7.251	0.1379	0.0096	0.0696	104.184	14.368	11.228	161.319	34
35	7.686	0.1301	0.0090	0.0690	111.435	14.498	11.432	165.743	35
36	8.147	0.1227	0.0084	0.0684	119.121	14.621	11.630	170.039	36
37	8.636	0.1158	0.0079	0.0679	127.268	14.737	11.821	174.207	37
38	9.154	0.1092	0.0074	0.0674	135.904	14.846	12.007	178.249	38
39	9.704	0.1031	0.0069	0.0669	145.058	14.949	12.186	182.165	39
40	10.286	0.0972	0.0065	0.0665	154.762	15.046	12.359	185.957	40
41	10.903	0.0917	0.0061	0.0661	165.048	15.138	12.526	189.626	41
42	11.557	0.0865	0.0057	0.0657	175.951	15.225	12.688	193.173	42
43	12.250	0.0816	0.0053	0.0653	187.508	15.306	12.845	196.602	43
44	12.985	0.0770	0.0050	0.0650	199.758	15.383	12.996	199.913	44
45	13.765	0.0727	0.0047	0.0647	212.744	15.456	13.141	203.110	45
46	14.590	0.0685	0.0044	0.0644	226.508	15.524	13.282	206.194	46
47	15.466	0.0647	0.0041	0.0641	241.099	15.589	13.418	209.168	47
48	16.394	0.0610	0.0039	0.0639	256.565	15.650	13.549	212.035	48
49	17.378	0.0575	0.0037	0.0637	272.958	15.708	13.675	214.797	49
50	18.420	0.0543	0.0034	0.0634	290.336	15.762	13.796	217.457	50

7%				Compound Interest Factors					7%
Period	Single Payment		Uniform Payment Series				Arithmetic Gradient		Period
	Compound Amount Factor	Present Value Factor	Sinking Fund Factor	Capital Recovery Factor	Compound Amount Factor	Present Value Factor	Gradient Uniform Series	Gradient Present Value	
	Find F Given P	Find P Given F	Find A Given F	Find A Given P	Find F Given A	Find P Given A	Find A Given G	Find P Given G	
n	F/P	P/F	A/F	A/P	F/A	P/A	A/G	P/G	*n*
1	1.070	0.9346	1.0000	1.0700	1.000	0.935	0.000	0.000	1
2	1.145	0.8734	0.4831	0.5531	2.070	1.808	0.483	0.873	2
3	1.225	0.8163	0.3111	0.3811	3.215	2.624	0.955	2.506	3
4	1.311	0.7629	0.2252	0.2952	4.440	3.387	1.416	4.795	4
5	1.403	0.7130	0.1739	0.2439	5.751	4.100	1.865	7.647	5
6	1.501	0.6663	0.1398	0.2098	7.153	4.767	2.303	10.978	6
7	1.606	0.6227	0.1156	0.1856	8.654	5.389	2.730	14.715	7
8	1.718	0.5820	0.0975	0.1675	10.260	5.971	3.147	18.789	8
9	1.838	0.5439	0.0835	0.1535	11.978	6.515	3.552	23.140	9
10	1.967	0.5083	0.0724	0.1424	13.816	7.024	3.946	27.716	10
11	2.105	0.4751	0.0634	0.1334	15.784	7.499	4.330	32.466	11
12	2.252	0.4440	0.0559	0.1259	17.888	7.943	4.703	37.351	12
13	2.410	0.4150	0.0497	0.1197	20.141	8.358	5.065	42.330	13
14	2.579	0.3878	0.0443	0.1143	22.550	8.745	5.417	47.372	14
15	2.759	0.3624	0.0398	0.1098	25.129	9.108	5.758	52.446	15
16	2.952	0.3387	0.0359	0.1059	27.888	9.447	6.090	57.527	16
17	3.159	0.3166	0.0324	0.1024	30.840	9.763	6.411	62.592	17
18	3.380	0.2959	0.0294	0.0994	33.999	10.059	6.722	67.622	18
19	3.617	0.2765	0.0268	0.0968	37.379	10.336	7.024	72.599	19
20	3.870	0.2584	0.0244	0.0944	40.995	10.594	7.316	77.509	20
21	4.141	0.2415	0.0223	0.0923	44.865	10.836	7.599	82.339	21
22	4.430	0.2257	0.0204	0.0904	49.006	11.061	7.872	87.079	22
23	4.741	0.2109	0.0187	0.0887	53.436	11.272	8.137	91.720	23
24	5.072	0.1971	0.0172	0.0872	58.177	11.469	8.392	96.255	24
25	5.427	0.1842	0.0158	0.0858	63.249	11.654	8.639	100.676	25
26	5.807	0.1722	0.0146	0.0846	68.676	11.826	8.877	104.981	26
27	6.214	0.1609	0.0134	0.0834	74.484	11.987	9.107	109.166	27
28	6.649	0.1504	0.0124	0.0824	80.698	12.137	9.329	113.226	28
29	7.114	0.1406	0.0114	0.0814	87.347	12.278	9.543	117.162	29
30	7.612	0.1314	0.0106	0.0806	94.461	12.409	9.749	120.972	30
31	8.145	0.1228	0.0098	0.0798	102.073	12.532	9.947	124.655	31
32	8.715	0.1147	0.0091	0.0791	110.218	12.647	10.138	128.212	32
33	9.325	0.1072	0.0084	0.0784	118.933	12.754	10.322	131.643	33
34	9.978	0.1002	0.0078	0.0778	128.259	12.854	10.499	134.951	34
35	10.677	0.0937	0.0072	0.0772	138.237	12.948	10.669	138.135	35
36	11.424	0.0875	0.0067	0.0767	148.913	13.035	10.832	141.199	36
37	12.224	0.0818	0.0062	0.0762	160.337	13.117	10.989	144.144	37
38	13.079	0.0765	0.0058	0.0758	172.561	13.193	11.140	146.973	38
39	13.995	0.0715	0.0054	0.0754	185.640	13.265	11.285	149.688	39
40	14.974	0.0668	0.0050	0.0750	199.635	13.332	11.423	152.293	40
41	16.023	0.0624	0.0047	0.0747	214.610	13.394	11.557	154.789	41
42	17.144	0.0583	0.0043	0.0743	230.632	13.452	11.684	157.181	42
43	18.344	0.0545	0.0040	0.0740	247.776	13.507	11.807	159.470	43
44	19.628	0.0509	0.0038	0.0738	266.121	13.558	11.924	161.661	44
45	21.002	0.0476	0.0035	0.0735	285.749	13.606	12.036	163.756	45
46	22.473	0.0445	0.0033	0.0733	306.752	13.650	12.143	165.758	46
47	24.046	0.0416	0.0030	0.0730	329.224	13.692	12.246	167.671	47
48	25.729	0.0389	0.0028	0.0728	353.270	13.730	12.345	169.498	48
49	27.530	0.0363	0.0026	0.0726	378.999	13.767	12.439	171.242	49
50	29.457	0.0339	0.0025	0.0725	406.529	13.801	12.529	172.905	50

8%	Compound Interest Factors							8%
Period	Single Payment		Uniform Payment Series				Arithmetic Gradient		Period
	Compound Amount Factor	Present Value Factor	Sinking Fund Factor	Capital Recovery Factor	Compound Amount Factor	Present Value Factor	Gradient Uniform Series	Gradient Present Value	
	Find F Given P	Find P Given F	Find A Given F	Find A Given P	Find F Given A	Find P Given A	Find A Given G	Find P Given G	
n	F/P	P/F	A/F	A/P	F/A	P/A	A/G	P/G	n
1	1.080	0.9259	1.0000	1.0800	1.000	0.926	0.000	0.000	1
2	1.166	0.8573	0.4808	0.5608	2.080	1.783	0.481	0.857	2
3	1.260	0.7938	0.3080	0.3880	3.246	2.577	0.949	2.445	3
4	1.360	0.7350	0.2219	0.3019	4.506	3.312	1.404	4.650	4
5	1.469	0.6806	0.1705	0.2505	5.867	3.993	1.846	7.372	5
6	1.587	0.6302	0.1363	0.2163	7.336	4.623	2.276	10.523	6
7	1.714	0.5835	0.1121	0.1921	8.923	5.206	2.694	14.024	7
8	1.851	0.5403	0.0940	0.1740	10.637	5.747	3.099	17.806	8
9	1.999	0.5002	0.0801	0.1601	12.488	6.247	3.491	21.808	9
10	2.159	0.4632	0.0690	0.1490	14.487	6.710	3.871	25.977	10
11	2.332	0.4289	0.0601	0.1401	16.645	7.139	4.240	30.266	11
12	2.518	0.3971	0.0527	0.1327	18.977	7.536	4.596	34.634	12
13	2.720	0.3677	0.0465	0.1265	21.495	7.904	4.940	39.046	13
14	2.937	0.3405	0.0413	0.1213	24.215	8.244	5.273	43.472	14
15	3.172	0.3152	0.0368	0.1168	27.152	8.559	5.594	47.886	15
16	3.426	0.2919	0.0330	0.1130	30.324	8.851	5.905	52.264	16
17	3.700	0.2703	0.0296	0.1096	33.750	9.122	6.204	56.588	17
18	3.996	0.2502	0.0267	0.1067	37.450	9.372	6.492	60.843	18
19	4.316	0.2317	0.0241	0.1041	41.446	9.604	6.770	65.013	19
20	4.661	0.2145	0.0219	0.1019	45.762	9.818	7.037	69.090	20
21	5.034	0.1987	0.0198	0.0998	50.423	10.017	7.294	73.063	21
22	5.437	0.1839	0.0180	0.0980	55.457	10.201	7.541	76.926	22
23	5.871	0.1703	0.0164	0.0964	60.893	10.371	7.779	80.673	23
24	6.341	0.1577	0.0150	0.0950	66.765	10.529	8.007	84.300	24
25	6.848	0.1460	0.0137	0.0937	73.106	10.675	8.225	87.804	25
26	7.396	0.1352	0.0125	0.0925	79.954	10.810	8.435	91.184	26
27	7.988	0.1252	0.0114	0.0914	87.351	10.935	8.636	94.439	27
28	8.627	0.1159	0.0105	0.0905	95.339	11.051	8.829	97.569	28
29	9.317	0.1073	0.0096	0.0896	103.966	11.158	9.013	100.574	29
30	10.063	0.0994	0.0088	0.0888	113.283	11.258	9.190	103.456	30
31	10.868	0.0920	0.0081	0.0881	123.346	11.350	9.358	106.216	31
32	11.737	0.0852	0.0075	0.0875	134.214	11.435	9.520	108.857	32
33	12.676	0.0789	0.0069	0.0869	145.951	11.514	9.674	111.382	33
34	13.690	0.0730	0.0063	0.0863	158.627	11.587	9.821	113.792	34
35	14.785	0.0676	0.0058	0.0858	172.317	11.655	9.961	116.092	35
36	15.968	0.0626	0.0053	0.0853	187.102	11.717	10.095	118.284	36
37	17.246	0.0580	0.0049	0.0849	203.070	11.775	10.222	120.371	37
38	18.625	0.0537	0.0045	0.0845	220.316	11.829	10.344	122.358	38
39	20.115	0.0497	0.0042	0.0842	238.941	11.879	10.460	124.247	39
40	21.725	0.0460	0.0039	0.0839	259.057	11.925	10.570	126.042	40
41	23.462	0.0426	0.0036	0.0836	280.781	11.967	10.675	127.747	41
42	25.339	0.0395	0.0033	0.0833	304.244	12.007	10.774	129.365	42
43	27.367	0.0365	0.0030	0.0830	329.583	12.043	10.869	130.900	43
44	29.556	0.0338	0.0028	0.0828	356.950	12.077	10.959	132.355	44
45	31.920	0.0313	0.0026	0.0826	386.506	12.108	11.045	133.733	45
46	34.474	0.0290	0.0024	0.0824	418.426	12.137	11.126	135.038	46
47	37.232	0.0269	0.0022	0.0822	452.900	12.164	11.203	136.274	47
48	40.211	0.0249	0.0020	0.0820	490.132	12.189	11.276	137.443	48
49	43.427	0.0230	0.0019	0.0819	530.343	12.212	11.345	138.548	49
50	46.902	0.0213	0.0017	0.0817	573.770	12.233	11.411	139.593	50

9%	Compound Interest Factors								9%
Period	Single Payment		Uniform Payment Series				Arithmetic Gradient		Period
	Compound Amount Factor	Present Value Factor	Sinking Fund Factor	Capital Recovery Factor	Compound Amount Factor	Present Value Factor	Gradient Uniform Series	Gradient Present Value	
	Find F Given P	Find P Given F	Find A Given F	Find A Given P	Find F Given A	Find P Given A	Find A Given G	Find P Given G	
n	F/P	P/F	A/F	A/P	F/A	P/A	A/G	P/G	n
1	1.090	0.9174	1.0000	1.0900	1.000	0.917	0.000	0.000	1
2	1.188	0.8417	0.4785	0.5685	2.090	1.759	0.478	0.842	2
3	1.295	0.7722	0.3051	0.3951	3.278	2.531	0.943	2.386	3
4	1.412	0.7084	0.2187	0.3087	4.573	3.240	1.393	4.511	4
5	1.539	0.6499	0.1671	0.2571	5.985	3.890	1.828	7.111	5
6	1.677	0.5963	0.1329	0.2229	7.523	4.486	2.250	10.092	6
7	1.828	0.5470	0.1087	0.1987	9.200	5.033	2.657	13.375	7
8	1.993	0.5019	0.0907	0.1807	11.028	5.535	3.051	16.888	8
9	2.172	0.4604	0.0768	0.1668	13.021	5.995	3.431	20.571	9
10	2.367	0.4224	0.0658	0.1558	15.193	6.418	3.798	24.373	10
11	2.580	0.3875	0.0569	0.1469	17.560	6.805	4.151	28.248	11
12	2.813	0.3555	0.0497	0.1397	20.141	7.161	4.491	32.159	12
13	3.066	0.3262	0.0436	0.1336	22.953	7.487	4.818	36.073	13
14	3.342	0.2992	0.0384	0.1284	26.019	7.786	5.133	39.963	14
15	3.642	0.2745	0.0341	0.1241	29.361	8.061	5.435	43.807	15
16	3.970	0.2519	0.0303	0.1203	33.003	8.313	5.724	47.585	16
17	4.328	0.2311	0.0270	0.1170	36.974	8.544	6.002	51.282	17
18	4.717	0.2120	0.0242	0.1142	41.301	8.756	6.269	54.886	18
19	5.142	0.1945	0.0217	0.1117	46.018	8.950	6.524	58.387	19
20	5.604	0.1784	0.0195	0.1095	51.160	9.129	6.767	61.777	20
21	6.109	0.1637	0.0176	0.1076	56.765	9.292	7.001	65.051	21
22	6.659	0.1502	0.0159	0.1059	62.873	9.442	7.223	68.205	22
23	7.258	0.1378	0.0144	0.1044	69.532	9.580	7.436	71.236	23
24	7.911	0.1264	0.0130	0.1030	76.790	9.707	7.638	74.143	24
25	8.623	0.1160	0.0118	0.1018	84.701	9.823	7.832	76.926	25
26	9.399	0.1064	0.0107	0.1007	93.324	9.929	8.016	79.586	26
27	10.245	0.0976	0.0097	0.0997	102.723	10.027	8.191	82.124	27
28	11.167	0.0895	0.0089	0.0989	112.968	10.116	8.357	84.542	28
29	12.172	0.0822	0.0081	0.0981	124.135	10.198	8.515	86.842	29
30	13.268	0.0754	0.0073	0.0973	136.308	10.274	8.666	89.028	30
31	14.462	0.0691	0.0067	0.0967	149.575	10.343	8.808	91.102	31
32	15.763	0.0634	0.0061	0.0961	164.037	10.406	8.944	93.069	32
33	17.182	0.0582	0.0056	0.0956	179.800	10.464	9.072	94.931	33
34	18.728	0.0534	0.0051	0.0951	196.982	10.518	9.193	96.693	34
35	20.414	0.0490	0.0046	0.0946	215.711	10.567	9.308	98.359	35
36	22.251	0.0449	0.0042	0.0942	236.125	10.612	9.417	99.932	36
37	24.254	0.0412	0.0039	0.0939	258.376	10.653	9.520	101.416	37
38	26.437	0.0378	0.0035	0.0935	282.630	10.691	9.617	102.816	38
39	28.816	0.0347	0.0032	0.0932	309.066	10.726	9.709	104.135	39
40	31.409	0.0318	0.0030	0.0930	337.882	10.757	9.796	105.376	40
41	34.236	0.0292	0.0027	0.0927	369.292	10.787	9.878	106.545	41
42	37.318	0.0268	0.0025	0.0925	403.528	10.813	9.955	107.643	42
43	40.676	0.0246	0.0023	0.0923	440.846	10.838	10.027	108.676	43
44	44.337	0.0226	0.0021	0.0921	481.522	10.861	10.096	109.646	44
45	48.327	0.0207	0.0019	0.0919	525.859	10.881	10.160	110.556	45
46	52.677	0.0190	0.0017	0.0917	574.186	10.900	10.221	111.410	46
47	57.418	0.0174	0.0016	0.0916	626.863	10.918	10.278	112.211	47
48	62.585	0.0160	0.0015	0.0915	684.280	10.934	10.332	112.962	48
49	68.218	0.0147	0.0013	0.0913	746.866	10.948	10.382	113.666	49
50	74.358	0.0134	0.0012	0.0912	815.084	10.962	10.430	114.325	50

10%				Compound Interest Factors					10%
Period	Single Payment		Uniform Payment Series				Arithmetic Gradient		Period
	Compound Amount Factor	Present Value Factor	Sinking Fund Factor	Capital Recovery Factor	Compound Amount Factor	Present Value Factor	Gradient Uniform Series	Gradient Present Value	
	Find F Given P	Find P Given F	Find A Given F	Find A Given P	Find F Given A	Find P Given A	Find A Given G	Find P Given G	
n	F/P	P/F	A/F	A/P	F/A	P/A	A/G	P/G	n
1	1.100	0.9091	1.0000	1.1000	1.000	0.909	0.000	0.000	1
2	1.210	0.8264	0.4762	0.5762	2.100	1.736	0.476	0.826	2
3	1.331	0.7513	0.3021	0.4021	3.310	2.487	0.937	2.329	3
4	1.464	0.6830	0.2155	0.3155	4.641	3.170	1.381	4.378	4
5	1.611	0.6209	0.1638	0.2638	6.105	3.791	1.810	6.862	5
6	1.772	0.5645	0.1296	0.2296	7.716	4.355	2.224	9.684	6
7	1.949	0.5132	0.1054	0.2054	9.487	4.868	2.622	12.763	7
8	2.144	0.4665	0.0874	0.1874	11.436	5.335	3.004	16.029	8
9	2.358	0.4241	0.0736	0.1736	13.579	5.759	3.372	19.421	9
10	2.594	0.3855	0.0627	0.1627	15.937	6.145	3.725	22.891	10
11	2.853	0.3505	0.0540	0.1540	18.531	6.495	4.064	26.396	11
12	3.138	0.3186	0.0468	0.1468	21.384	6.814	4.388	29.901	12
13	3.452	0.2897	0.0408	0.1408	24.523	7.103	4.699	33.377	13
14	3.797	0.2633	0.0357	0.1357	27.975	7.367	4.996	36.800	14
15	4.177	0.2394	0.0315	0.1315	31.772	7.606	5.279	40.152	15
16	4.595	0.2176	0.0278	0.1278	35.950	7.824	5.549	43.416	16
17	5.054	0.1978	0.0247	0.1247	40.545	8.022	5.807	46.582	17
18	5.560	0.1799	0.0219	0.1219	45.599	8.201	6.053	49.640	18
19	6.116	0.1635	0.0195	0.1195	51.159	8.365	6.286	52.583	19
20	6.727	0.1486	0.0175	0.1175	57.275	8.514	6.508	55.407	20
21	7.400	0.1351	0.0156	0.1156	64.002	8.649	6.719	58.110	21
22	8.140	0.1228	0.0140	0.1140	71.403	8.772	6.919	60.689	22
23	8.954	0.1117	0.0126	0.1126	79.543	8.883	7.108	63.146	23
24	9.850	0.1015	0.0113	0.1113	88.497	8.985	7.288	65.481	24
25	10.835	0.0923	0.0102	0.1102	98.347	9.077	7.458	67.696	25
26	11.918	0.0839	0.0092	0.1092	109.182	9.161	7.619	69.794	26
27	13.110	0.0763	0.0083	0.1083	121.100	9.237	7.770	71.777	27
28	14.421	0.0693	0.0075	0.1075	134.210	9.307	7.914	73.650	28
29	15.863	0.0630	0.0067	0.1067	148.631	9.370	8.049	75.415	29
30	17.449	0.0573	0.0061	0.1061	164.494	9.427	8.176	77.077	30
31	19.194	0.0521	0.0055	0.1055	181.943	9.479	8.296	78.640	31
32	21.114	0.0474	0.0050	0.1050	201.138	9.526	8.409	80.108	32
33	23.225	0.0431	0.0045	0.1045	222.252	9.569	8.515	81.486	33
34	25.548	0.0391	0.0041	0.1041	245.477	9.609	8.615	82.777	34
35	28.102	0.0356	0.0037	0.1037	271.024	9.644	8.709	83.987	35
36	30.913	0.0323	0.0033	0.1033	299.127	9.677	8.796	85.119	36
37	34.004	0.0294	0.0030	0.1030	330.039	9.706	8.879	86.178	37
38	37.404	0.0267	0.0027	0.1027	364.043	9.733	8.956	87.167	38
39	41.145	0.0243	0.0025	0.1025	401.448	9.757	9.029	88.091	39
40	45.259	0.0221	0.0023	0.1023	442.593	9.779	9.096	88.953	40
41	49.785	0.0201	0.0020	0.1020	487.852	9.799	9.160	89.756	41
42	54.764	0.0183	0.0019	0.1019	537.637	9.817	9.219	90.505	42
43	60.240	0.0166	0.0017	0.1017	592.401	9.834	9.274	91.202	43
44	66.264	0.0151	0.0015	0.1015	652.641	9.849	9.326	91.851	44
45	72.890	0.0137	0.0014	0.1014	718.905	9.863	9.374	92.454	45
46	80.180	0.0125	0.0013	0.1013	791.795	9.875	9.419	93.016	46
47	88.197	0.0113	0.0011	0.1011	871.975	9.887	9.461	93.537	47
48	97.017	0.0103	0.0010	0.1010	960.172	9.897	9.500	94.022	48
49	106.719	0.0094	0.0009	0.1009	1057.190	9.906	9.537	94.471	49
50	117.391	0.0085	0.0009	0.1009	1163.909	9.915	9.570	94.889	50

12%			Compound Interest Factors				Arithmetic Gradient		12%
Period	Single Payment		Uniform Payment Series				Arithmetic Gradient		Period
	Compound Amount Factor	Present Value Factor	Sinking Fund Factor	Capital Recovery Factor	Compound Amount Factor	Present Value Factor	Gradient Uniform Series	Gradient Present Value	
	Find F Given P	Find P Given F	Find A Given F	Find A Given P	Find F Given A	Find P Given A	Find A Given G	Find P Given G	
n	F/P	P/F	A/F	A/P	F/A	P/A	A/G	P/G	*n*
1	1.120	0.8929	1.0000	1.1200	1.000	0.893	0.000	0.000	1
2	1.254	0.7972	0.4717	0.5917	2.120	1.690	0.472	0.797	2
3	1.405	0.7118	0.2963	0.4163	3.374	2.402	0.925	2.221	3
4	1.574	0.6355	0.2092	0.3292	4.779	3.037	1.359	4.127	4
5	1.762	0.5674	0.1574	0.2774	6.353	3.605	1.775	6.397	5
6	1.974	0.5066	0.1232	0.2432	8.115	4.111	2.172	8.930	6
7	2.211	0.4523	0.0991	0.2191	10.089	4.564	2.551	11.644	7
8	2.476	0.4039	0.0813	0.2013	12.300	4.968	2.913	14.471	8
9	2.773	0.3606	0.0677	0.1877	14.776	5.328	3.257	17.356	9
10	3.106	0.3220	0.0570	0.1770	17.549	5.650	3.585	20.254	10
11	3.479	0.2875	0.0484	0.1684	20.655	5.938	3.895	23.129	11
12	3.896	0.2567	0.0414	0.1614	24.133	6.194	4.190	25.952	12
13	4.363	0.2292	0.0357	0.1557	28.029	6.424	4.468	28.702	13
14	4.887	0.2046	0.0309	0.1509	32.393	6.628	4.732	31.362	14
15	5.474	0.1827	0.0268	0.1468	37.280	6.811	4.980	33.920	15
16	6.130	0.1631	0.0234	0.1434	42.753	6.974	5.215	36.367	16
17	6.866	0.1456	0.0205	0.1405	48.884	7.120	5.435	38.697	17
18	7.690	0.1300	0.0179	0.1379	55.750	7.250	5.643	40.908	18
19	8.613	0.1161	0.0158	0.1358	63.440	7.366	5.838	42.998	19
20	9.646	0.1037	0.0139	0.1339	72.052	7.469	6.020	44.968	20
21	10.804	0.0926	0.0122	0.1322	81.699	7.562	6.191	46.819	21
22	12.100	0.0826	0.0108	0.1308	92.503	7.645	6.351	48.554	22
23	13.552	0.0738	0.0096	0.1296	104.603	7.718	6.501	50.178	23
24	15.179	0.0659	0.0085	0.1285	118.155	7.784	6.641	51.693	24
25	17.000	0.0588	0.0075	0.1275	133.334	7.843	6.771	53.105	25
26	19.040	0.0525	0.0067	0.1267	150.334	7.896	6.892	54.418	26
27	21.325	0.0469	0.0059	0.1259	169.374	7.943	7.005	55.637	27
28	23.884	0.0419	0.0052	0.1252	190.699	7.984	7.110	56.767	28
29	26.750	0.0374	0.0047	0.1247	214.583	8.022	7.207	57.814	29
30	29.960	0.0334	0.0041	0.1241	241.333	8.055	7.297	58.782	30
31	33.555	0.0298	0.0037	0.1237	271.293	8.085	7.381	59.676	31
32	37.582	0.0266	0.0033	0.1233	304.848	8.112	7.459	60.501	32
33	42.092	0.0238	0.0029	0.1229	342.429	8.135	7.530	61.261	33
34	47.143	0.0212	0.0026	0.1226	384.521	8.157	7.596	61.961	34
35	52.800	0.0189	0.0023	0.1223	431.663	8.176	7.658	62.605	35
36	59.136	0.0169	0.0021	0.1221	484.463	8.192	7.714	63.197	36
37	66.232	0.0151	0.0018	0.1218	543.599	8.208	7.766	63.741	37
38	74.180	0.0135	0.0016	0.1216	609.831	8.221	7.814	64.239	38
39	83.081	0.0120	0.0015	0.1215	684.010	8.233	7.858	64.697	39
40	93.051	0.0107	0.0013	0.1213	767.091	8.244	7.899	65.116	40
41	104.217	0.0096	0.0012	0.1212	860.142	8.253	7.936	65.500	41
42	116.723	0.0086	0.0010	0.1210	964.359	8.262	7.970	65.851	42
43	130.730	0.0076	0.0009	0.1209	1081.083	8.270	8.002	66.172	43
44	146.418	0.0068	0.0008	0.1208	1211.813	8.276	8.031	66.466	44
45	163.988	0.0061	0.0007	0.1207	1358.230	8.283	8.057	66.734	45
46	183.666	0.0054	0.0007	0.1207	1522.218	8.288	8.082	66.979	46
47	205.706	0.0049	0.0006	0.1206	1705.884	8.293	8.104	67.203	47
48	230.391	0.0043	0.0005	0.1205	1911.590	8.297	8.124	67.407	48
49	258.038	0.0039	0.0005	0.1205	2141.981	8.301	8.143	67.593	49
50	289.002	0.0035	0.0004	0.1204	2400.018	8.304	8.160	67.762	50

14%					Compound Interest Factors				14%
Period	Single Payment		Uniform Payment Series				Arithmetic Gradient		Period
	Compound Amount Factor	Present Value Factor	Sinking Fund Factor	Capital Recovery Factor	Compound Amount Factor	Present Value Factor	Gradient Uniform Series	Gradient Present Value	
	Find F Given P	Find P Given F	Find A Given F	Find A Given P	Find F Given A	Find P Given A	Find A Given G	Find P Given G	
n	F/P	P/F	A/F	A/P	F/A	P/A	A/G	P/G	n
1	1.140	0.8772	1.0000	1.1400	1.000	0.877	0.000	0.000	1
2	1.300	0.7695	0.4673	0.6073	2.140	1.647	0.467	0.769	2
3	1.482	0.6750	0.2907	0.4307	3.440	2.322	0.913	2.119	3
4	1.689	0.5921	0.2032	0.3432	4.921	2.914	1.337	3.896	4
5	1.925	0.5194	0.1513	0.2913	6.610	3.433	1.740	5.973	5
6	2.195	0.4556	0.1172	0.2572	8.536	3.889	2.122	8.251	6
7	2.502	0.3996	0.0932	0.2332	10.730	4.288	2.483	10.649	7
8	2.853	0.3506	0.0756	0.2156	13.233	4.639	2.825	13.103	8
9	3.252	0.3075	0.0622	0.2022	16.085	4.946	3.146	15.563	9
10	3.707	0.2697	0.0517	0.1917	19.337	5.216	3.449	17.991	10
11	4.226	0.2366	0.0434	0.1834	23.045	5.453	3.733	20.357	11
12	4.818	0.2076	0.0367	0.1767	27.271	5.660	4.000	22.640	12
13	5.492	0.1821	0.0312	0.1712	32.089	5.842	4.249	24.825	13
14	6.261	0.1597	0.0266	0.1666	37.581	6.002	4.482	26.901	14
15	7.138	0.1401	0.0228	0.1628	43.842	6.142	4.699	28.862	15
16	8.137	0.1229	0.0196	0.1596	50.980	6.265	4.901	30.706	16
17	9.276	0.1078	0.0169	0.1569	59.118	6.373	5.089	32.430	17
18	10.575	0.0946	0.0146	0.1546	68.394	6.467	5.263	34.038	18
19	12.056	0.0829	0.0127	0.1527	78.969	6.550	5.424	35.531	19
20	13.743	0.0728	0.0110	0.1510	91.025	6.623	5.573	36.914	20
21	15.668	0.0638	0.0095	0.1495	104.768	6.687	5.711	38.190	21
22	17.861	0.0560	0.0083	0.1483	120.436	6.743	5.838	39.366	22
23	20.362	0.0491	0.0072	0.1472	138.297	6.792	5.955	40.446	23
24	23.212	0.0431	0.0063	0.1463	158.659	6.835	6.062	41.437	24
25	26.462	0.0378	0.0055	0.1455	181.871	6.873	6.161	42.344	25
26	30.167	0.0331	0.0048	0.1448	208.333	6.906	6.251	43.173	26
27	34.390	0.0291	0.0042	0.1442	238.499	6.935	6.334	43.929	27
28	39.204	0.0255	0.0037	0.1437	272.889	6.961	6.410	44.618	28
29	44.693	0.0224	0.0032	0.1432	312.094	6.983	6.479	45.244	29
30	50.950	0.0196	0.0028	0.1428	356.787	7.003	6.542	45.813	30
31	58.083	0.0172	0.0025	0.1425	407.737	7.020	6.600	46.330	31
32	66.215	0.0151	0.0021	0.1421	465.820	7.035	6.652	46.798	32
33	75.485	0.0132	0.0019	0.1419	532.035	7.048	6.700	47.222	33
34	86.053	0.0116	0.0016	0.1416	607.520	7.060	6.743	47.605	34
35	98.100	0.0102	0.0014	0.1414	693.573	7.070	6.782	47.952	35
36	111.834	0.0089	0.0013	0.1413	791.673	7.079	6.818	48.265	36
37	127.491	0.0078	0.0011	0.1411	903.507	7.087	6.850	48.547	37
38	145.340	0.0069	0.0010	0.1410	1030.998	7.094	6.880	48.802	38
39	165.687	0.0060	0.0009	0.1409	1176.338	7.100	6.906	49.031	39
40	188.884	0.0053	0.0007	0.1407	1342.025	7.105	6.930	49.238	40
41	215.327	0.0046	0.0007	0.1407	1530.909	7.110	6.952	49.423	41
42	245.473	0.0041	0.0006	0.1406	1746.236	7.114	6.971	49.590	42
43	279.839	0.0036	0.0005	0.1405	1991.709	7.117	6.989	49.741	43
44	319.017	0.0031	0.0004	0.1404	2271.548	7.120	7.004	49.875	44
45	363.679	0.0027	0.0004	0.1404	2590.565	7.123	7.019	49.996	45
46	414.594	0.0024	0.0003	0.1403	2954.244	7.126	7.032	50.105	46
47	472.637	0.0021	0.0003	0.1403	3368.838	7.128	7.043	50.202	47
48	538.807	0.0019	0.0003	0.1403	3841.475	7.130	7.054	50.289	48
49	614.239	0.0016	0.0002	0.1402	4380.282	7.131	7.063	50.368	49
50	700.233	0.0014	0.0002	0.1402	4994.521	7.133	7.071	50.438	50

16%	Compound Interest Factors								16%
Period	Single Payment		Uniform Payment Series				Arithmetic Gradient		Period
	Compound Amount Factor	Present Value Factor	Sinking Fund Factor	Capital Recovery Factor	Compound Amount Factor	Present Value Factor	Gradient Uniform Series	Gradient Present Value	
	Find F Given P	Find P Given F	Find A Given F	Find A Given P	Find F Given A	Find P Given A	Find A Given G	Find P Given G	
n	F/P	P/F	A/F	A/P	F/A	P/A	A/G	P/G	n
1	1.160	0.8621	1.0000	1.1600	1.000	0.862	0.000	0.000	1
2	1.346	0.7432	0.4630	0.6230	2.160	1.605	0.463	0.743	2
3	1.561	0.6407	0.2853	0.4453	3.506	2.246	0.901	2.024	3
4	1.811	0.5523	0.1974	0.3574	5.066	2.798	1.316	3.681	4
5	2.100	0.4761	0.1454	0.3054	6.877	3.274	1.706	5.586	5
6	2.436	0.4104	0.1114	0.2714	8.977	3.685	2.073	7.638	6
7	2.826	0.3538	0.0876	0.2476	11.414	4.039	2.417	9.761	7
8	3.278	0.3050	0.0702	0.2302	14.240	4.344	2.739	11.896	8
9	3.803	0.2630	0.0571	0.2171	17.519	4.607	3.039	14.000	9
10	4.411	0.2267	0.0469	0.2069	21.321	4.833	3.319	16.040	10
11	5.117	0.1954	0.0389	0.1989	25.733	5.029	3.578	17.994	11
12	5.936	0.1685	0.0324	0.1924	30.850	5.197	3.819	19.847	12
13	6.886	0.1452	0.0272	0.1872	36.786	5.342	4.041	21.590	13
14	7.988	0.1252	0.0229	0.1829	43.672	5.468	4.246	23.217	14
15	9.266	0.1079	0.0194	0.1794	51.660	5.575	4.435	24.728	15
16	10.748	0.0930	0.0164	0.1764	60.925	5.668	4.609	26.124	16
17	12.468	0.0802	0.0140	0.1740	71.673	5.749	4.768	27.407	17
18	14.463	0.0691	0.0119	0.1719	84.141	5.818	4.913	28.583	18
19	16.777	0.0596	0.0101	0.1701	98.603	5.877	5.046	29.656	19
20	19.461	0.0514	0.0087	0.1687	115.380	5.929	5.167	30.632	20
21	22.574	0.0443	0.0074	0.1674	134.841	5.973	5.277	31.518	21
22	26.186	0.0382	0.0064	0.1664	157.415	6.011	5.377	32.320	22
23	30.376	0.0329	0.0054	0.1654	183.601	6.044	5.467	33.044	23
24	35.236	0.0284	0.0047	0.1647	213.978	6.073	5.549	33.697	24
25	40.874	0.0245	0.0040	0.1640	249.214	6.097	5.623	34.284	25
26	47.414	0.0211	0.0034	0.1634	290.088	6.118	5.690	34.811	26
27	55.000	0.0182	0.0030	0.1630	337.502	6.136	5.750	35.284	27
28	63.800	0.0157	0.0025	0.1625	392.503	6.152	5.804	35.707	28
29	74.009	0.0135	0.0022	0.1622	456.303	6.166	5.853	36.086	29
30	85.850	0.0116	0.0019	0.1619	530.312	6.177	5.896	36.423	30
31	99.586	0.0100	0.0016	0.1616	616.162	6.187	5.936	36.725	31
32	115.520	0.0087	0.0014	0.1614	715.747	6.196	5.971	36.993	32
33	134.003	0.0075	0.0012	0.1612	831.267	6.203	6.002	37.232	33
34	155.443	0.0064	0.0010	0.1610	965.270	6.210	6.030	37.444	34
35	180.314	0.0055	0.0009	0.1609	1120.713	6.215	6.055	37.633	35
36	209.164	0.0048	0.0008	0.1608	1301.027	6.220	6.077	37.800	36
37	242.631	0.0041	0.0007	0.1607	1510.191	6.224	6.097	37.948	37
38	281.452	0.0036	0.0006	0.1606	1752.822	6.228	6.115	38.080	38
39	326.484	0.0031	0.0005	0.1605	2034.273	6.231	6.130	38.196	39
40	378.721	0.0026	0.0004	0.1604	2360.757	6.233	6.144	38.299	40
41	439.317	0.0023	0.0004	0.1604	2739.478	6.236	6.156	38.390	41
42	509.607	0.0020	0.0003	0.1603	3178.795	6.238	6.167	38.471	42
43	591.144	0.0017	0.0003	0.1603	3688.402	6.239	6.177	38.542	43
44	685.727	0.0015	0.0002	0.1602	4279.546	6.241	6.186	38.605	44
45	795.444	0.0013	0.0002	0.1602	4965.274	6.242	6.193	38.660	45
46	922.715	0.0011	0.0002	0.1602	5760.718	6.243	6.200	38.709	46
47	1070.349	0.0009	0.0001	0.1601	6683.433	6.244	6.206	38.752	47
48	1241.605	0.0008	0.0001	0.1601	7753.782	6.245	6.211	38.789	48
49	1440.262	0.0007	0.0001	0.1601	8995.387	6.246	6.216	38.823	49
50	1670.704	0.0006	0.0001	0.1601	10435.649	6.246	6.220	38.852	50

18%			Compound Interest Factors						18%
Period	Single Payment		Uniform Payment Series				Arithmetic Gradient		Period
	Compound Amount Factor	Present Value Factor	Sinking Fund Factor	Capital Recovery Factor	Compound Amount Factor	Present Value Factor	Gradient Uniform Series	Gradient Present Value	
	Find F Given P	Find P Given F	Find A Given F	Find A Given P	Find F Given A	Find P Given A	Find A Given G	Find P Given G	
n	F/P	P/F	A/F	A/P	F/A	P/A	A/G	P/G	n
1	1.180	0.8475	1.0000	1.1800	1.000	0.847	0.000	0.000	1
2	1.392	0.7182	0.4587	0.6387	2.180	1.566	0.459	0.718	2
3	1.643	0.6086	0.2799	0.4599	3.572	2.174	0.890	1.935	3
4	1.939	0.5158	0.1917	0.3717	5.215	2.690	1.295	3.483	4
5	2.288	0.4371	0.1398	0.3198	7.154	3.127	1.673	5.231	5
6	2.700	0.3704	0.1059	0.2859	9.442	3.498	2.025	7.083	6
7	3.185	0.3139	0.0824	0.2624	12.142	3.812	2.353	8.967	7
8	3.759	0.2660	0.0652	0.2452	15.327	4.078	2.656	10.829	8
9	4.435	0.2255	0.0524	0.2324	19.086	4.303	2.936	12.633	9
10	5.234	0.1911	0.0425	0.2225	23.521	4.494	3.194	14.352	10
11	6.176	0.1619	0.0348	0.2148	28.755	4.656	3.430	15.972	11
12	7.288	0.1372	0.0286	0.2086	34.931	4.793	3.647	17.481	12
13	8.599	0.1163	0.0237	0.2037	42.219	4.910	3.845	18.877	13
14	10.147	0.0985	0.0197	0.1997	50.818	5.008	4.025	20.158	14
15	11.974	0.0835	0.0164	0.1964	60.965	5.092	4.189	21.327	15
16	14.129	0.0708	0.0137	0.1937	72.939	5.162	4.337	22.389	16
17	16.672	0.0600	0.0115	0.1915	87.068	5.222	4.471	23.348	17
18	19.673	0.0508	0.0096	0.1896	103.740	5.273	4.592	24.212	18
19	23.214	0.0431	0.0081	0.1881	123.414	5.316	4.700	24.988	19
20	27.393	0.0365	0.0068	0.1868	146.628	5.353	4.798	25.681	20
21	32.324	0.0309	0.0057	0.1857	174.021	5.384	4.885	26.300	21
22	38.142	0.0262	0.0048	0.1848	206.345	5.410	4.963	26.851	22
23	45.008	0.0222	0.0041	0.1841	244.487	5.432	5.033	27.339	23
24	53.109	0.0188	0.0035	0.1835	289.494	5.451	5.095	27.772	24
25	62.669	0.0160	0.0029	0.1829	342.603	5.467	5.150	28.155	25
26	73.949	0.0135	0.0025	0.1825	405.272	5.480	5.199	28.494	26
27	87.260	0.0115	0.0021	0.1821	479.221	5.492	5.243	28.791	27
28	102.967	0.0097	0.0018	0.1818	566.481	5.502	5.281	29.054	28
29	121.501	0.0082	0.0015	0.1815	669.447	5.510	5.315	29.284	29
30	143.371	0.0070	0.0013	0.1813	790.948	5.517	5.345	29.486	30
31	169.177	0.0059	0.0011	0.1811	934.319	5.523	5.371	29.664	31
32	199.629	0.0050	0.0009	0.1809	1103.496	5.528	5.394	29.819	32
33	235.563	0.0042	0.0008	0.1808	1303.125	5.532	5.415	29.955	33
34	277.964	0.0036	0.0006	0.1806	1538.688	5.536	5.433	30.074	34
35	327.997	0.0030	0.0006	0.1806	1816.652	5.539	5.449	30.177	35
36	387.037	0.0026	0.0005	0.1805	2144.649	5.541	5.462	30.268	36
37	456.703	0.0022	0.0004	0.1804	2531.686	5.543	5.474	30.347	37
38	538.910	0.0019	0.0003	0.1803	2988.389	5.545	5.485	30.415	38
39	635.914	0.0016	0.0003	0.1803	3527.299	5.547	5.494	30.475	39
40	750.378	0.0013	0.0002	0.1802	4163.213	5.548	5.502	30.527	40
41	885.446	0.0011	0.0002	0.1802	4913.591	5.549	5.509	30.572	41
42	1044.827	0.0010	0.0002	0.1802	5799.038	5.550	5.515	30.611	42
43	1232.896	0.0008	0.0001	0.1801	6843.865	5.551	5.521	30.645	43
44	1454.817	0.0007	0.0001	0.1801	8076.760	5.552	5.525	30.675	44
45	1716.684	0.0006	0.0001	0.1801	9531.577	5.552	5.529	30.701	45
46	2025.687	0.0005	0.0001	0.1801	11248.261	5.553	5.533	30.723	46
47	2390.311	0.0004	0.0001	0.1801	13273.948	5.553	5.536	30.742	47
48	2820.567	0.0004	0.0001	0.1801	15664.259	5.554	5.539	30.759	48
49	3328.269	0.0003	0.0001	0.1801	18484.825	5.554	5.541	30.773	49
50	3927.357	0.0003	0.0000	0.1800	21813.094	5.554	5.543	30.786	50

20%				Compound Interest Factors					20%
Period	Single Payment		Uniform Payment Series				Arithmetic Gradient		Period
	Compound Amount Factor	Present Value Factor	Sinking Fund Factor	Capital Recovery Factor	Compound Amount Factor	Present Value Factor	Gradient Uniform Series	Gradient Present Value	
	Find *F* Given *P*	Find *P* Given *F*	Find *A* Given *F*	Find *A* Given *P*	Find *F* Given *A*	Find *P* Given *A*	Find *A* Given *G*	Find *P* Given *G*	
n	*F/P*	*P/F*	*A/F*	*A/P*	*F/A*	*P/A*	*A/G*	*P/G*	*n*
1	1.200	0.8333	1.0000	1.2000	1.000	0.833	0.000	0.000	1
2	1.440	0.6944	0.4545	0.6545	2.200	1.528	0.455	0.694	2
3	1.728	0.5787	0.2747	0.4747	3.640	2.106	0.879	1.852	3
4	2.074	0.4823	0.1863	0.3863	5.368	2.589	1.274	3.299	4
5	2.488	0.4019	0.1344	0.3344	7.442	2.991	1.641	4.906	5
6	2.986	0.3349	0.1007	0.3007	9.930	3.326	1.979	6.581	6
7	3.583	0.2791	0.0774	0.2774	12.916	3.605	2.290	8.255	7
8	4.300	0.2326	0.0606	0.2606	16.499	3.837	2.576	9.883	8
9	5.160	0.1938	0.0481	0.2481	20.799	4.031	2.836	11.434	9
10	6.192	0.1615	0.0385	0.2385	25.959	4.192	3.074	12.887	10
11	7.430	0.1346	0.0311	0.2311	32.150	4.327	3.289	14.233	11
12	8.916	0.1122	0.0253	0.2253	39.581	4.439	3.484	15.467	12
13	10.699	0.0935	0.0206	0.2206	48.497	4.533	3.660	16.588	13
14	12.839	0.0779	0.0169	0.2169	59.196	4.611	3.817	17.601	14
15	15.407	0.0649	0.0139	0.2139	72.035	4.675	3.959	18.509	15
16	18.488	0.0541	0.0114	0.2114	87.442	4.730	4.085	19.321	16
17	22.186	0.0451	0.0094	0.2094	105.931	4.775	4.198	20.042	17
18	26.623	0.0376	0.0078	0.2078	128.117	4.812	4.298	20.680	18
19	31.948	0.0313	0.0065	0.2065	154.740	4.843	4.386	21.244	19
20	38.338	0.0261	0.0054	0.2054	186.688	4.870	4.464	21.739	20
21	46.005	0.0217	0.0044	0.2044	225.026	4.891	4.533	22.174	21
22	55.206	0.0181	0.0037	0.2037	271.031	4.909	4.594	22.555	22
23	66.247	0.0151	0.0031	0.2031	326.237	4.925	4.647	22.887	23
24	79.497	0.0126	0.0025	0.2025	392.484	4.937	4.694	23.176	24
25	95.396	0.0105	0.0021	0.2021	471.981	4.948	4.735	23.428	25
26	114.475	0.0087	0.0018	0.2018	567.377	4.956	4.771	23.646	26
27	137.371	0.0073	0.0015	0.2015	681.853	4.964	4.802	23.835	27
28	164.845	0.0061	0.0012	0.2012	819.223	4.970	4.829	23.999	28
29	197.814	0.0051	0.0010	0.2010	984.068	4.975	4.853	24.141	29
30	237.376	0.0042	0.0008	0.2008	1181.882	4.979	4.873	24.263	30
31	284.852	0.0035	0.0007	0.2007	1419.258	4.982	4.891	24.368	31
32	341.822	0.0029	0.0006	0.2006	1704.109	4.985	4.906	24.459	32
33	410.186	0.0024	0.0005	0.2005	2045.931	4.988	4.919	24.537	33
34	492.224	0.0020	0.0004	0.2004	2456.118	4.990	4.931	24.604	34
35	590.668	0.0017	0.0003	0.2003	2948.341	4.992	4.941	24.661	35
36	708.802	0.0014	0.0003	0.2003	3539.009	4.993	4.949	24.711	36
37	850.562	0.0012	0.0002	0.2002	4247.811	4.994	4.956	24.753	37
38	1020.675	0.0010	0.0002	0.2002	5098.373	4.995	4.963	24.789	38
39	1224.810	0.0008	0.0002	0.2002	6119.048	4.996	4.968	24.820	39
40	1469.772	0.0007	0.0001	0.2001	7343.858	4.997	4.973	24.847	40
41	1763.726	0.0006	0.0001	0.2001	8813.629	4.997	4.977	24.870	41
42	2116.471	0.0005	0.0001	0.2001	10577.355	4.998	4.980	24.889	42
43	2539.765	0.0004	0.0001	0.2001	12693.826	4.998	4.983	24.906	43
44	3047.718	0.0003	0.0001	0.2001	15233.592	4.998	4.986	24.920	44
45	3657.262	0.0003	0.0001	0.2001	18281.310	4.999	4.988	24.932	45
46	4388.714	0.0002	0.0000	0.2000	21938.572	4.999	4.990	24.942	46
47	5266.457	0.0002	0.0000	0.2000	26327.286	4.999	4.991	24.951	47
48	6319.749	0.0002	0.0000	0.2000	31593.744	4.999	4.992	24.958	48
49	7583.698	0.0001	0.0000	0.2000	37913.492	4.999	4.994	24.964	49
50	9100.438	0.0001	0.0000	0.2000	45497.191	4.999	4.995	24.970	50

25%				Compound Interest Factors					25%
Period	Single Payment		Uniform Payment Series				Arithmetic Gradient		Period
	Compound Amount Factor	Present Value Factor	Sinking Fund Factor	Capital Recovery Factor	Compound Amount Factor	Present Value Factor	Gradient Uniform Series	Gradient Present Value	
	Find F Given P	Find P Given F	Find A Given F	Find A Given P	Find F Given A	Find P Given A	Find A Given G	Find P Given G	
n	F/P	P/F	A/F	A/P	F/A	P/A	A/G	P/G	n
1	1.250	0.8000	1.0000	1.2500	1.000	0.800	0.000	0.000	1
2	1.563	0.6400	0.4444	0.6944	2.250	1.440	0.444	0.640	2
3	1.953	0.5120	0.2623	0.5123	3.813	1.952	0.852	1.664	3
4	2.441	0.4096	0.1734	0.4234	5.766	2.362	1.225	2.893	4
5	3.052	0.3277	0.1218	0.3718	8.207	2.689	1.563	4.204	5
6	3.815	0.2621	0.0888	0.3388	11.259	2.951	1.868	5.514	6
7	4.768	0.2097	0.0663	0.3163	15.073	3.161	2.142	6.773	7
8	5.960	0.1678	0.0504	0.3004	19.842	3.329	2.387	7.947	8
9	7.451	0.1342	0.0388	0.2888	25.802	3.463	2.605	9.021	9
10	9.313	0.1074	0.0301	0.2801	33.253	3.571	2.797	9.987	10
11	11.642	0.0859	0.0235	0.2735	42.566	3.656	2.966	10.846	11
12	14.552	0.0687	0.0184	0.2684	54.208	3.725	3.115	11.602	12
13	18.190	0.0550	0.0145	0.2645	68.760	3.780	3.244	12.262	13
14	22.737	0.0440	0.0115	0.2615	86.949	3.824	3.356	12.833	14
15	28.422	0.0352	0.0091	0.2591	109.687	3.859	3.453	13.326	15
16	35.527	0.0281	0.0072	0.2572	138.109	3.887	3.537	13.748	16
17	44.409	0.0225	0.0058	0.2558	173.636	3.910	3.608	14.108	17
18	55.511	0.0180	0.0046	0.2546	218.045	3.928	3.670	14.415	18
19	69.389	0.0144	0.0037	0.2537	273.556	3.942	3.722	14.674	19
20	86.736	0.0115	0.0029	0.2529	342.945	3.954	3.767	14.893	20
21	108.420	0.0092	0.0023	0.2523	429.681	3.963	3.805	15.078	21
22	135.525	0.0074	0.0019	0.2519	538.101	3.970	3.836	15.233	22
23	169.407	0.0059	0.0015	0.2515	673.626	3.976	3.863	15.362	23
24	211.758	0.0047	0.0012	0.2512	843.033	3.981	3.886	15.471	24
25	264.698	0.0038	0.0009	0.2509	1054.791	3.985	3.905	15.562	25
26	330.872	0.0030	0.0008	0.2508	1319.489	3.988	3.921	15.637	26
27	413.590	0.0024	0.0006	0.2506	1650.361	3.990	3.935	15.700	27
28	516.988	0.0019	0.0005	0.2505	2063.952	3.992	3.946	15.752	28
29	646.235	0.0015	0.0004	0.2504	2580.939	3.994	3.955	15.796	29
30	807.794	0.0012	0.0003	0.2503	3227.174	3.995	3.963	15.832	30
31	1009.742	0.0010	0.0002	0.2502	4034.968	3.996	3.969	15.861	31
32	1262.177	0.0008	0.0002	0.2502	5044.710	3.997	3.975	15.886	32
33	1577.722	0.0006	0.0002	0.2502	6306.887	3.997	3.979	15.906	33
34	1972.152	0.0005	0.0001	0.2501	7884.609	3.998	3.983	15.923	34
35	2465.190	0.0004	0.0001	0.2501	9856.761	3.998	3.986	15.937	35
36	3081.488	0.0003	0.0001	0.2501	12321.952	3.999	3.988	15.948	36
37	3851.860	0.0003	0.0001	0.2501	15403.440	3.999	3.990	15.957	37
38	4814.825	0.0002	0.0001	0.2501	19255.299	3.999	3.992	15.965	38
39	6018.531	0.0002	0.0000	0.2500	24070.124	3.999	3.994	15.971	39
40	7523.164	0.0001	0.0000	0.2500	30088.655	3.999	3.995	15.977	40
41	9403.955	0.0001	0.0000	0.2500	37611.819	4.000	3.996	15.981	41
42	11754.944	0.0001	0.0000	0.2500	47015.774	4.000	3.996	15.984	42
43	14693.679	0.0001	0.0000	0.2500	58770.718	4.000	3.997	15.987	43
44	18367.099	0.0001	0.0000	0.2500	73464.397	4.000	3.998	15.990	44
45	22958.874	0.0000	0.0000	0.2500	91831.496	4.000	3.998	15.991	45
46	28698.593	0.0000	0.0000	0.2500	114790.370	4.000	3.998	15.993	46
47	35873.241	0.0000	0.0000	0.2500	143488.963	4.000	3.999	15.994	47
48	44841.551	0.0000	0.0000	0.2500	179362.203	4.000	3.999	15.995	48
49	56051.939	0.0000	0.0000	0.2500	224203.754	4.000	3.999	15.996	49
50	70064.923	0.0000	0.0000	0.2500	280255.693	4.000	3.999	15.997	50

30%				Compound Interest Factors					30%
Period	Single Payment		Uniform Payment Series				Arithmetic Gradient		Period
	Compound Amount Factor	Present Value Factor	Sinking Fund Factor	Capital Recovery Factor	Compound Amount Factor	Present Value Factor	Gradient Uniform Series	Gradient Present Value	
	Find *F* Given *P*	Find *P* Given *F*	Find *A* Given *F*	Find *A* Given *P*	Find *F* Given *A*	Find *P* Given *A*	Find *A* Given *G*	Find *P* Given *G*	
n	*F/P*	*P/F*	*A/F*	*A/P*	*F/A*	*P/A*	*A/G*	*P/G*	*n*
1	1.300	0.7692	1.0000	1.3000	1.000	0.769	0.000	0.000	1
2	1.690	0.5917	0.4348	0.7348	2.300	1.361	0.435	0.592	2
3	2.197	0.4552	0.2506	0.5506	3.990	1.816	0.827	1.502	3
4	2.856	0.3501	0.1616	0.4616	6.187	2.166	1.178	2.552	4
5	3.713	0.2693	0.1106	0.4106	9.043	2.436	1.490	3.630	5
6	4.827	0.2072	0.0784	0.3784	12.756	2.643	1.765	4.666	6
7	6.275	0.1594	0.0569	0.3569	17.583	2.802	2.006	5.622	7
8	8.157	0.1226	0.0419	0.3419	23.858	2.925	2.216	6.480	8
9	10.604	0.0943	0.0312	0.3312	32.015	3.019	2.396	7.234	9
10	13.786	0.0725	0.0235	0.3235	42.619	3.092	2.551	7.887	10
11	17.922	0.0558	0.0177	0.3177	56.405	3.147	2.683	8.445	11
12	23.298	0.0429	0.0135	0.3135	74.327	3.190	2.795	8.917	12
13	30.288	0.0330	0.0102	0.3102	97.625	3.223	2.889	9.314	13
14	39.374	0.0254	0.0078	0.3078	127.913	3.249	2.969	9.644	14
15	51.186	0.0195	0.0060	0.3060	167.286	3.268	3.034	9.917	15
16	66.542	0.0150	0.0046	0.3046	218.472	3.283	3.089	10.143	16
17	86.504	0.0116	0.0035	0.3035	285.014	3.295	3.135	10.328	17
18	112.455	0.0089	0.0027	0.3027	371.518	3.304	3.172	10.479	18
19	146.192	0.0068	0.0021	0.3021	483.973	3.311	3.202	10.602	19
20	190.050	0.0053	0.0016	0.3016	630.165	3.316	3.228	10.702	20
21	247.065	0.0040	0.0012	0.3012	820.215	3.320	3.248	10.783	21
22	321.184	0.0031	0.0009	0.3009	1067.280	3.323	3.265	10.848	22
23	417.539	0.0024	0.0007	0.3007	1388.464	3.325	3.278	10.901	23
24	542.801	0.0018	0.0006	0.3006	1806.003	3.327	3.289	10.943	24
25	705.641	0.0014	0.0004	0.3004	2348.803	3.329	3.298	10.977	25
26	917.333	0.0011	0.0003	0.3003	3054.444	3.330	3.305	11.005	26
27	1192.533	0.0008	0.0003	0.3003	3971.778	3.331	3.311	11.026	27
28	1550.293	0.0006	0.0002	0.3002	5164.311	3.331	3.315	11.044	28
29	2015.381	0.0005	0.0001	0.3001	6714.604	3.332	3.319	11.058	29
30	2619.996	0.0004	0.0001	0.3001	8729.985	3.332	3.322	11.069	30
31	3405.994	0.0003	0.0001	0.3001	11349.981	3.332	3.324	11.078	31
32	4427.793	0.0002	0.0001	0.3001	14755.975	3.333	3.326	11.085	32
33	5756.130	0.0002	0.0001	0.3001	19183.768	3.333	3.328	11.090	33
34	7482.970	0.0001	0.0000	0.3000	24939.899	3.333	3.329	11.094	34
35	9727.860	0.0001	0.0000	0.3000	32422.868	3.333	3.330	11.098	35
36	12646.219	0.0001	0.0000	0.3000	42150.729	3.333	3.330	11.101	36
37	16440.084	0.0001	0.0000	0.3000	54796.947	3.333	3.331	11.103	37
38	21372.109	0.0000	0.0000	0.3000	71237.031	3.333	3.332	11.105	38
39	27783.742	0.0000	0.0000	0.3000	92609.141	3.333	3.332	11.106	39
40	36118.865	0.0000	0.0000	0.3000	120392.883	3.333	3.332	11.107	40
41	46954.524	0.0000	0.0000	0.3000	156511.748	3.333	3.332	11.108	41
42	61040.882	0.0000	0.0000	0.3000	203466.272	3.333	3.333	11.109	42
43	79353.146	0.0000	0.0000	0.3000	264507.153	3.333	3.333	11.109	43
44	103159.090	0.0000	0.0000	0.3000	343860.299	3.333	3.333	11.110	44
45	134106.817	0.0000	0.0000	0.3000	447019.389	3.333	3.333	11.110	45
46	174338.862	0.0000	0.0000	0.3000	581126.206	3.333	3.333	11.110	46
47	226640.520	0.0000	0.0000	0.3000	755465.067	3.333	3.333	11.110	47
48	294632.676	0.0000	0.0000	0.3000	982105.588	3.333	3.333	11.111	48
49	383022.479	0.0000	0.0000	0.3000	1276738.264	3.333	3.333	11.111	49
50	497929.223	0.0000	0.0000	0.3000	1659760.743	3.333	3.333	11.111	50

Index